Finite element modelling of composite materials and structures

F L Matthews, G A O Davies,
D Hitchings and C Soutis

CRC Press
Boca Raton Boston New York Washington, DC

WOODHEAD PUBLISHING LIMITED

Cambridge England

Published by Woodhead Publishing Limited, Abington Hall, Abington
Cambridge CB21 6AH, England
www.woodheadpublishing.com

Published in North America by CRC Press LLC, 6000 Broken Sound Parkway, NW,
Suite 300, Boca Raton, FL 33487, USA

First published 2000, Woodhead Publishing Ltd and CRC Press LLC
Reprinted 2003, 2004, 2007

This edition is for sale in Indian Subcontinent only.

British Library Cataloguing in Publication Data
A catalogue record for this book is available from the British Library.

Library of Congress Cataloging in Publication Data
A catalog record for this book is available from the Library of Congress.

Woodhead Publishing ISBN-13: 978-1-85573-422-7
Woodhead Publishing ISBN-10: 1-85573-422-2
CRC Press ISBN-13: 978-0-8493-0846-8
CRC Press ISBN-10: 0-8493-0846-1
CRC Press order number: WP0846

Printed in India by Replika Press Pvt Ltd

Contents

iv   Contents

Part IV Analytical and numerical modelling
C. SOUTIS

Preface

The Centre for Composite Materials at Imperial College has for many years organised a series of training courses on various aspects of composite materials. In 1996, recognising the increasing interest in structural analysis, a new course on 'Finite Element Analysis of Composites' was launched. The course was given jointly by the Centre and the Department of Aeronautics, and supported by the Engineering & Physical Sciences Research Council (EPSRC) as an MSc level module.

This book is based on the lecture notes prepared for the above course. The emphasis throughout is on long fibre-reinforced polymer (FRP) matrix composites, although any general analysis would be applicable to other forms of composite. The book starts with a review of the basic behaviour of FRP. Then, the fundamentals of finite element (FE) analysis are rehearsed. Following this, special issues relating to the applications of FE to FRP are discussed. Finally, a number of particular situations, such as holes and free edges, are presented, with FE results set alongside classical analysis and experimental data.

Authors

The authors of this text are all from the Imperial College of Science Technology & Medicine, University of London.

Prof. F. L. Matthews   Director, Centre for Composite Materials
Prof. G. A. O. Davies   Senior Research Fellow & Senior Research Investigator, Department of Aeronautics
Mr D. Hitchings   Senior Lecturer, Department of Aeronautics
Dr C. Soutis   Reader, Department of Aeronautics

Acknowledgement

The authors, and especially F. L. Matthews in his editorial role, are indebted to Mrs Sabine Grune von Stieglitz for her invaluable contribution in assembling this book from a multitude of electronic and paper sources.

Part I
Overview and review of composite materials

F. L. MATTHEWS

This part serves merely to 'set the scene' for the remainder of the book and also to revise the fundamentals of composite materials. The basic nature of fibre-reinforced plastic and its constituents is reviewed. Following this, two-dimensional stress–strain analysis is covered, leading to laminated plate theory. Finally some limitations of the latter are discussed.

Overview

1.1 Composite materials

1.1.1 General

The group of materials known as 'composites' is extremely large, although its boundaries depend on definition. Basically, we can consider a composite as any material that is a combination of two or more distinct constituents. This definition would encompass bricks, concrete, wood, bone, as well as modern synthetic composites such as fibre-reinforced plastics (FRP). The latter have become increasingly important over the past 50 years, and are now the first choice for fabricating structures where low weight in combination with high strength and stiffness are required. Such materials are sometimes referred to as 'high-performance' composites, and would often be composed of carbon fibres and epoxy resin.

Of course, composites can also be made by combining fibres with a metal or a ceramic matrix, but at the moment these have a very small market share and would be considered as specialist materials.

1.1.2 Properties and applications

Density, stiffness (modulus) and strength are the properties that initially come to mind when thinking of FRP, and these would certainly be the design drivers for materials' selection for transport applications such as aircraft, motor vehicles and trains. However, this is a very narrow view of the potential of such materials, and they often score over metals and other conventional materials because of other mechanical, physical or chemical properties.

For example, FRPs are extremely corrosion-resistant and are, in consequence, used for chemical plant, water transport and storage, and flue gas desulphurisation plant. They have interesting electromagnetic properties and because of this glass fibre-reinforced resin is used to construct mine counter measures naval vessels, the requirement being a non-magnetic material. As another illustration, carbon fibre-reinforced epoxy is used in medical applications because it is transparent to X-rays. Thermal properties can also be important; such materials have a low conductivity, making them useful for fire protection, and can have a zero coefficient of expansion, hence making possible the construction of temperature-stable components.

A key issue is that of cost. Although the constituents of FRP are relatively expensive, compared with conventional materials, the final component may well be less costly than one fabricated from metal, say. This is due, in part, to having a more integrated construction (lower 'parts count'), but depends crucially on the level of automation in the manufacturing process.

1.1.3 Production methods

The important factor about FRP is that, unlike metals, the material is made at the same time as the component. While this gives an increased freedom to the design process, it does mean that designers/analysts must pay close attention to the fabrication of the component or structure. The same constituents if processed by one method, could produce a composite with some properties modified compared with those produced by another method. Also, it is important to realise that the designer may call for a particular configuration that cannot actually be fabricated; for example, it is not possible to maintain a constant fibre angle and wall thickness along the length of a tapered cylinder produced by filament winding. Good communication at all levels throughout the whole design/fabrication/evaluation process is, therefore, essential.

1.2   Structural analysis

1.2.1 Classical analysis

The use of classical (continuum) methods of stress analysis has developed over many decades to give techniques that can be applied satisfactorily to a vast range of situations. Such analyses are based on the application of the equations of equilibrium and compatibility, together with the stress–strain relations for the material, to produce governing equations which must be solved to obtain displacements and stresses. Usually, assumptions must be made before a solution can be effected. So, for example, problems are considered as one- or two-dimensional, as when considering beams and plates, respectively. Often we take the material to be isotropic, but many analyses also exist for anisotropic materials.

As we move away from simple situations, say from a plain rectangular plate to one containing a cut-out, the governing equations become increasingly complicated and require ever-more sophisticated mathematical techniques to solve them. Classical methods are limited to simple geometries and 'real' structural features, e.g. the details of attachment of a stringer to a skin panel, cannot be analysed. In such cases we have to resort to finite element methods.

1.2.2 Finite element analysis

Finite element (FE) analysis is merely an alternative approach to solving the governing equations of a structural problem. Hence, FE and classical methods will produce identical results for the same problem, provided the former method is correctly applied.

The method consists of imagining the structure to be composed of discrete parts (i.e. finite elements), which are then assembled in such a way as to represent the distortion of the structure under the specified loads. Each element has an assumed displacement field, and part of the skill of applying the method is in selecting appropriate elements of the correct size and distributions (the FE 'mesh').

The FE method was initially developed for isotropic materials and the majority of elements available (the 'library') in any software package would be for such materials. To apply the technique to composites requires different element formulations that adequately represent their anisotropic, or orthotropic, stiffness and strength, as well as the laminated form of construction often used.

The main purpose of this book is to discuss the special issues that must be considered when using the FE method to analyse composite materials and structures. One of the key factors is the low through-thickness stiffness and strength; the former results in the need to adapt elements used for plates, and the latter results in the need to model delamination. Also some features of composite construction, such as filament winding, cannot easily be represented (if at all) by some FE packages. Following a review of composites and the FE method, the application of the method to composites is discussed in detail. The particular issues are then illustrated via a number of examples taken from particular situations.

2.1 Basic characteristics

2.1.1 Definitions and classification

A composite is a mixture of two or more distinct constituents or phases. In addition three other criteria are normally satisfied before we call a material a composite. Firstly, both constituents have to be present in reasonable proportions. Secondly, the constituent phases should have distinctly different properties, such that the composite's properties are noticeably different from the properties of the constituents. Lastly, a synthetic composite is usually produced by deliberately mixing and combining the constituents by various means.

We know that composites have two (or more) chemically distinct phases on a microscopic scale, separated by a distinct interface, and it is important to be able to specify these constituents. The constituent that is continuous and is often, but not always, present in the greater quantity in the composite is termed the matrix. The normal view is that it is the properties of the matrix that are improved upon when incorporating another constituent to produce a composite. A composite may have a ceramic, metallic or polymeric matrix. The mechanical properties of these three classes of material differ considerably. As a generalisation, polymers have low strengths and Young's moduli, ceramics are strong, stiff and brittle, and metals have intermediate strengths and moduli, together with good ductilities, i.e. they are not brittle. Because of their economic importance, the emphasis in this text will be on polymer matrix composites (PMCs).

The second constituent is known to as the reinforcing phase, or reinforcement, as it enhances or reinforces the mechanical properties of the matrix. In most cases the reinforcement is harder, stronger and stiffer than the matrix, although there are some exceptions; for example, ductile metal reinforcement in a ceramic matrix and rubberlike reinforcement in a brittle polymer matrix. At least one of the dimensions of the reinforcement is

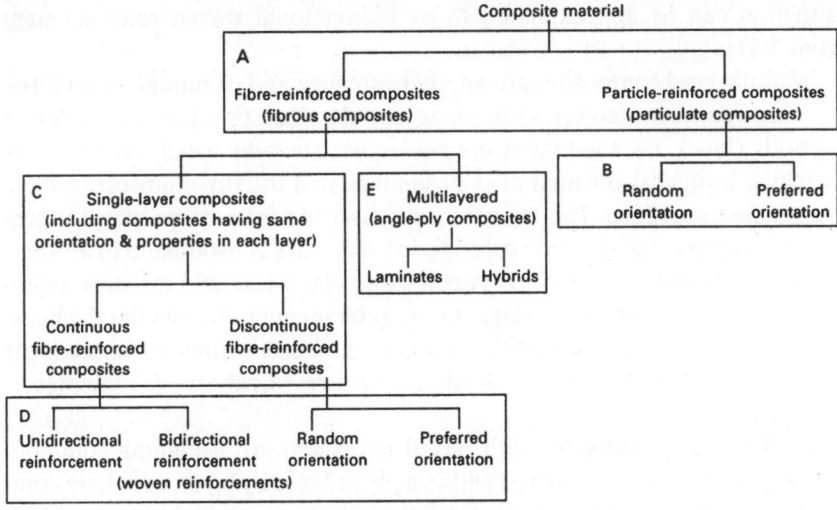

2.1 Classification of composite materials.

small, say less than 500 μm and sometimes only of the order of a micrometre. The geometry of the reinforcing phase is one of the major parameters in determining the effectiveness of the reinforcement; in other words, the mechanical properties of composites are a function of the shape and dimensions of the reinforcement. We usually describe the reinforcement as being either fibrous or particulate. Figure 2.1 represents a commonly employed classification scheme for composite materials which utilizes this designation for the reinforcement (Fig. 2.1, block A).

Particulate reinforcements have dimensions that are approximately equal in all directions. The shape of the reinforcing particles may be spherical, cubic, platelet or any regular or irregular geometry. The arrangement of the particulate reinforcement may be random or with a preferred orientation, and this characteristic is also used as a part of the classification scheme (block B). In the majority of particulate-reinforced composites the orientation of the particles is considered, for practical purposes, to be random.

A fibrous reinforcement is characterized by its length being much greater than its cross-section dimensions. However, the ratio of length to a cross-section dimension, known as the aspect ratio, can vary considerably. In single-layer composites long fibres with high aspect ratios give what are called continuous fibre-reinforced composites, whereas discontinuous fibre composites are fabricated using short fibres of low aspect ratio (block C). The orientation of the discontinuous fibres may be random or preferred. The frequently encountered preferred orientation in the case of a continuous fibre composite is termed unidirectional and the corresponding random situation can be approximated to by bidirectional woven reinforcement (block D).

Multilayered composites are another category, and commonly used form, of fibre-reinforced composites. These are classified as either laminates or hybrids (block E). Laminates are sheet constructions which are made by stacking layers (also called plies or laminae and usually unidirectional) in a specified sequence. The layers are often in the form of 'prepreg' (fibres pre-impregnated with partly cured resin) which are consolidated in an auto-clave. A laminate may have between 4 and 400 layers and the fibre orien-tation changes from layer to layer in a regular manner through the thickness of the laminate, e.g. a 0/90/0 stacking sequence results in a cross-ply composite. We shall refer in detail to the stress analysis of laminates in Chapter 3.

Hybrids are composites with mixed fibres and are becoming common-place. The fibres may be mixed within a ply or layer by layer, and these com-posites are designed to benefit from the different properties of the fibres employed. For example, a mixture of glass and carbon fibres incorporated into a polymer matrix gives a relatively inexpensive composite, owing to the low cost of glass fibres, but with mechanical properties enhanced by the excellent stiffness of carbon.

2.1.2 The matrix and reinforcement

Most composites are designed to exploit an improvement in mechanical properties. Even for composites produced essentially for their physical properties, the mechanical properties can play an important role during component manufacture and service. The strengths of fibres are generally much higher than those of their monolithic counterparts owing to the presence of defects in the latter.

There are of course many properties other than strength that we have to take into account when selecting a reinforcement. In the case of fibres the flexibility is important as it determines whether the fibres may be easily woven or not, and influences the choice of method for composite manu-facture. The flexibility of a fibre depends on Young's modulus and the diam-eter of the fibre, decreasing as diameter increases.

Clearly, single fibres, because of their small cross-section dimensions, are not directly usable in structural applications. This problem may be over-come by embedding the fibres in a material to hold the fibres apart, to protect the surface of the fibres, and to facilitate the production of compo-nents. The embedding material is the matrix. The amount of reinforcement that can be incorporated in a given matrix is limited by a number of factors. For example with particulate-reinforced metals the reinforcement con-tent is usually kept to less than 40 vol% (0.4 volume fraction) owing to processing difficulties and increasing brittleness at higher contents. On the other hand, the processing methods for fibre-reinforced polymers are capable of producing composites with a high proportion of fibres, and the upper limit of about 70 vol% (0.7 volume fraction) is set by the need to avoid fibre–fibre contact, which results in fibre damage.

Finally, the fact that the reinforcement is bonded to the matrix means that any loads applied to a composite are carried by both constituents. As in most cases the reinforcement is the stiffer and stronger constituent, it is the principal load-bearer. The matrix is said to have transferred the load to the reinforcement.

2.1.3 Factors that determine properties

The fabrication and properties of composites are strongly influenced by the proportions and properties of the matrix and the reinforcement. The proportions can be expressed either via the weight fraction (w), which is relevant to fabrication, or via the volume fraction (v), which is commonly used in property calculations.

The definitions of w and v are related simply to the ratios of weight (W) or volume (V) as shown below.

Volume fractions:

$$v_f = V_f/V_c \quad \text{and} \quad v_m = V_m/V_c \tag{2.1a}$$

Weight fractions:

$$w_f = W_f/W_c \quad \text{and} \quad w_m = W_m/W_c \tag{2.1b}$$

where the subscripts m, f and c refer to the matrix, fibre (or in the more general case, reinforcement) and composite respectively.

We note that:

$$v_f + v_m = 1$$

and

$$w_f + w_m = 1$$

We can relate weight to volume fractions by introducing the density, ρ, of the composite and its constituents.

We can show that:

$$\rho_c = \rho_f v_f + \rho_m v_m \tag{2.2}$$

and

$$\frac{1}{\rho_c} = \frac{w_f}{\rho_f} + \frac{w_m}{\rho_m} \tag{2.3}$$

Also we have:

$$w_f = \frac{W_f}{W_c} = \frac{\rho_f V_f}{\rho_c V_c} = \left(\frac{\rho_f}{\rho_c}\right) v_f$$

and similarly

$$w_m = \frac{W_m}{W_c} = \frac{\rho_m V_m}{\rho_c V_c} = \left(\frac{\rho_m}{\rho_c}\right) v_m \qquad [2.4]$$

We can see that we can convert from weight fraction to volume fraction, and vice versa, provided the densities of the reinforcement (ρ_f) and the matrix (ρ_m) are known.

The chemical and strength characteristics of the interface between the fibres and the matrix are particularly important in determining the properties of the composite. The interfacial bond strength has to be sufficient for load to be transferred from the matrix to the fibres if the composite is to be stronger than the unreinforced matrix. On the other hand, if we are also concerned with the toughness of the composite, the interface must not be so strong that it does not fail and allow toughening mechanisms such as debonding and fibre pull-out to take place.

Other parameters that may significantly affect the properties of a composite are the shape, size, orientation and distribution of the reinforcement, and various features of the matrix such as the grain size for polycrystalline matrices. These, together with volume fraction, constitute what is called the microstructure of the composite.

The volume fraction is generally regarded as the single most important parameter influencing the composite's properties. Also, it is an easily controllable manufacturing variable by which the properties of a composite may be altered to suit the application.

A problem encountered during manufacture is maintaining a uniform distribution of the reinforcement. Ideally, a composite should be homogeneous, or uniform, but this is difficult to achieve. Homogeneity is an important characteristic that determines the extent to which a representative volume of the material may differ in physical and mechanical properties from the average properties of the material. Non-uniformity of the system should be avoided as much as possible because it reduces those properties that are governed by the weakest part of the composite.

The orientation of the reinforcement within the matrix affects the isotropy of the system. When the reinforcement is in the form of equiaxial particles, the composite behaves essentially as an isotropic material whose properties are independent of direction. When the dimensions of the reinforcement are unequal, the composite can behave as if isotropic provided the reinforcement is randomly oriented, as in a randomly oriented short fibre-reinforced composite. In other cases the manufacturing process may

Table 2.1 Typical properties of some artificial fibres

	Density (Mg/m³) ρ	Young's modulus (GPa) E_f	Tensile strength (MPa) $\hat{\sigma}_{Tf}$	Failure strain (%)
E-glass	2.54	70	2200	3.1
Aramid (Kevlar 49)	1.45	130	2900	2.5
SiC (Nicalon)	2.60	250	2200	0.9
Alumina (FP)	3.90	380	1400	0.4
Boron	2.65	420	3500	0.8
Polyethylene (S1000)	0.97	172	2960	1.7
Carbon (HM)	1.86	380	2700	0.7

induce orientation of the reinforcement and hence loss of isotropy; the composite is then said to be anisotropic. In components manufactured from continuous fibre-reinforced composites, such as unidirectional or cross-ply laminates, anisotropy may be desirable as it can be arranged for the maximum service stress to be in the direction that has the highest strength. Indeed, a primary advantage of these composites is the ability to control the anisotropy of a component by design and fabrication.

2.2   Fibres and matrices

2.2.1   Fibres

We have seen that the reinforcement in a composite may be fibrous or particulate. A wide range of both these forms of reinforcement is available for use in the production of composite materials but most of the major developments in recent times have been in the area of fibrous reinforcement. The underlying philosophy in the design of fibre composite materials is to find or to make a fibre material of high elastic modulus and strength, and preferably low density, and then to arrange the fibres in a suitable manner to give useful engineering properties to the final product.

Many different fibres are manufactured for the reinforcement in composites and some typical properties are given in Table 2.1.[1] The values for stiffness and strength given in the table should be viewed with some caution. The manufacture of the fibres involves a number of processing steps and variability of properties from one fibre to another can be large, even when made by the same process. Between fibres of the same material made by different processes, the resulting microstructure and properties can differ even more markedly. Furthermore, the high tensile strength of freshly made fibres is normally reduced by surface damage caused during subsequent handling and storage. Finally, any variation in size leads to a range of strength values. Most fibres are brittle and show only elastic extension before fracture.

2.2.2 Synthetic organic fibres

The organic reinforcement market is dominated by aramid fibres, although there is increasing interest in polyethylene fibres.

2.2.2.1 Aramid

There are a number of commercially available aramid fibres, e.g. Kevlar (Du Pont), Twaron (Akzo) and Technora (Teijin); of these Kevlar is the most well known. The aramids can be viewed as nylon with extra benzene rings in the polymer chain to increase stiffness. Alternative nomenclature that the reader might encounter is aromatic polyamide or poly(phenylene teraphthalamide).

Aramid fibres have good high temperature properties for a polymeric material. They have a glass transition temperature of about 360 °C, burn with difficulty and do not melt like nylon. A loss in performance due to carburation occurs around 425 °C but they can be used at elevated temperatures for sustained periods and even at 300 °C for a limited time. Aramid fibres do have a tendency to degrade in sunlight and when exposed to moisture. Also, they have low compressive strength. Their dimensional stability is good as the coefficient of thermal expansion is low (approximately $-4 \times 10^{-6} \, K^{-1}$). Other properties that are important in certain applications are low electrical and thermal conductivity and high thermal capacity.

2.2.2.2 Polyethylene

In the late 1980s a number of polyethylene fibres became commercially available including Spectra (Allied Signal) and Dyneema (DSM). Polyethylene has the lowest density of any readily available fibre (about $0.97 \, Mg/m^3$) and hence its specific properties are good and superior to those of Kevlar. However, polyethylene has a low melting point of 135 °C and readily creeps at elevated temperature, so use is restricted to temperatures below 100 °C.

2.2.3 Synthetic inorganic fibres

2.2.3.1 Glass

Glass is a non-crystalline material with a short-range network structure. As such it has no distinctive microstructure and the properties, which are determined mainly by composition and surface finish, are isotropic. There are many groups of glasses, for example silica, oxynitride, phosphate and halide glasses, but from the point of view of composite technology only the silica glasses are currently of importance. However, even within this group of glasses the composition, and hence properties, vary considerably.

The most commonly used glass fibre is E-glass, the E being an abbreviation for 'electrical'. This glass is based on the ternary system $CaO–Al_2O_3–SiO_2$. Freshly drawn and carefully handled fibres have tensile strengths of approximately $E_f/20$ but a typical value may be nearer to $E_f/50$, where E_f is Young's modulus (typically about 70 GPa).

S-glass, known as R-glass in Europe, is based on the $SiO_2–Al_2O_3–MgO$ system. This fibre has higher stiffness and strength (hence the designation S) than E-glass. It also retains its improved properties to higher temperatures. However, it is more difficult to draw into fibres because of its limited working range and is therefore more expensive.

Although the performance of the widely used general-purpose E-glass is satisfactory in near neutral aqueous solutions, it is liable to degradation in environments which are highly acidic or alkaline.

2.2.3.2 Carbon

Carbon fibres are produced by many companies and the world production capacity exceeds 20000 tonnes. In spite of this large production capacity carbon fibres are still relatively expensive. Nevertheless the usage for carbon fibres continues to increase, and many companies have recently increased their production capacity.

The structure and properties of these fibres vary considerably and new fibres are always under test. For example two recent introductions are hollow fibres and coiled fibres. The former are designed to impart better impact toughness to carbon-reinforced polymers, whereas the latter are capable of extending many times their original length without loss of elasticity.

Carbon has two well-known crystalline forms (diamond and graphite) but it also exists in quasi-crystalline and glassy states. As far as fibre technology is concerned, graphite is the most important structural form of carbon. The graphitic structure consists of strongly bonded hexagonal layers, which are called the basal planes, with weak interlayer van der Waals bonds.

Alignment of the basal planes parallel to the fibre axis gives stiff fibres with relatively low density.

Graphite sublimes at 3700 °C but starts to oxidise in air at around 500 °C; carbon fibres can, however, be used at temperatures exceeding 2500 °C if protected from oxygen. Carbon is a good electrical conductor which, depending on the circumstances, can be advantageous or not. Carbon fibres are produced from a variety of precursors. The mechanical properties vary greatly with the precursor used and the processing conditions employed, as these determine the perfection and alignment of the crystals. The main precursors are polyacrylonitrile (PAN) and pitch.

An unusual characteristic of carbon fibres is their very low, or even slightly negative, coefficient of longitudinal expansion. As for other properties, the coefficient of thermal expansion depends on the fabrication route and hence degree of graphitization and crystal orientation. Ultrahigh modulus carbon fibres have negative coefficients of expansion of approximately $-1.4 \times 10^{-6} \mathrm{K}^{-1}$ and are employed in the production of polymer matrix composites with near-zero thermal expansion.

2.2.4 Polymer matrices

The most common matrix materials for composites are polymeric. The reasons for this are twofold. Firstly, in general the mechanical properties of polymers are inadequate for many structural purposes. In particular their strength and stiffness are low compared with metals and ceramics. This means that there is a considerable benefit to be gained by reinforcing polymers, and that the reinforcement does not have to have exceptional properties.

Secondly, and most importantly, the processing of polymer matrix composites (PMCs) need not involve high pressures and does not require high temperatures. It follows that problems associated with the degradation of the reinforcement during manufacture are less significant for PMCs than for composites with other matrices. Also the equipment required for PMCs may be simpler.

There are three classes of polymers, thermosets, thermoplastics and rubbers, all important as far as matrices of PMCs are concerned. Rubbers will not be dealt with here, although they constitute an important group, being widely used in reinforced form in motor car tyres, for example. Within any class there are many different polymers, e.g. epoxy, polyester, polyimide and phenolic are all thermosets. Even a given polymer, such as a polyester, exists in many forms; there are a large number of formulations, curing agents and fillers, which results in an extensive range of properties for polyesters. Indeed, polyesters, and other polymers, are often marketed according to their properties by the employment of descriptive terms including

Table 2.2 Some typical properties of polymer matrices

	Epoxy	Nylon (6.6)	Polycarbonate	Polyester
Density (Mg/m^3)	1.1–1.4	1.1	1.1–1.2	1.1–1.5
Young's modulus (GPa)	2.1–6.0	1.4–2.8	2.2–2.4	1.3–4.5
Tensile strength (MPa)	35–90	60–70	45–70	45–85
Ductility (%)		30–100	90–110	
Fracture toughness K_{IC} ($MPa\,m^{1/2}$) G_{IC} (kJ/m^2)	0.6–1.0 0.02			0.5
Thermal expansion ($10^{-6}\,K^{-1}$)	55–110	90		100–200
Glass transition temperature (°C)	120–190		150	
Melting point (°C)		261		

'general purpose', 'chemically resistant' and 'heat resistant'. Bearing in mind also the variety of materials used for reinforcement, and the possible arrangements of the reinforcement (random chopped fibres, unidirectional, numerous weaves, etc.) it is clear that the range of properties exhibited by PMCs is quite remarkable.

The main disadvantages of PMCs are their low maximum working temperatures, high coefficients of thermal expansion and hence dimensional instability, and sensitivity to radiation and moisture. The absorption of water from the environment may have many harmful effects which degrade mechanical performance, including swelling, formation of internal stresses and lowering of the glass transition temperature.

It has been estimated that over three-quarters of all matrices of PMCs are thermosetting polymers. Thermosetting polymers, or thermosets, are resins that readily cross-link during curing. Curing involves the application of heat and pressure, or the addition of a catalyst known as a curing agent or hardener.

Thermosets are brittle at room temperature and have low fracture toughness values. Also, because of the cross-links, thermosets cannot be reheated and reshaped; thermosets just degrade on reheating, and in some cases may burn, but do not soften sufficiently for reshaping. Thermosets may be used at higher temperatures as they have higher softening temperatures and better creep properties than thermoplastics. Finally, they are more resistant to chemical attack than most thermoplastics. Typical properties are given in Table 2.2.[1]

Thermoplastics readily flow under stress at elevated temperatures, so allowing them to be fabricated into the required component, and become solid and retain their shape when cooled to room temperature. These polymers may be repeatedly heated, fabricated and cooled and consequently scrap may be recycled, though there is evidence that this slightly degrades the properties, probably because of a reduction in molecular weight. Well-known thermoplastics include acrylic, nylon (polyamide), polystyrene, polyethylene, polypropylene and polyetheretherketone. Some typical properties are given in Table 2.2.

2.3   Summary

Polymer matrix composites (PMCs) are the best established form of advanced composite materials. Of the three classes of polymers used as matrices, thermosets, thermoplastics and rubbers, thermosets dominate the market for structural applications. The mechanical properties of PMCs can vary widely depending on the choice of fibre and matrix, the coating applied to the fibres, and the manufacturing route. The main reason for the popularity of PMCs is their ease of processing.

2.4   Reference

1 Matthews F L & Rawlings R D, *Composite Materials: Engineering and Science*, Cambridge, Woodhead, 1999.

3

Stiffness and strength of composites

3.1 Stiffness of unidirectional composites and laminates

3.1.1 Introduction

When analysing composite structures using finite elements we need to supply to the software appropriate input data for the material being used. Typically for composites we shall need moduli and strength of the single plies that constitute a laminate. Occasionally we may input the laminate properties directly.

In this chapter we shall address the essential background relating to the above issues. Initially we shall work with 'macromechanics' in which case we ignore the details of the fibres and matrix and their interactions. Later in the chapter we shall look at these interactions in what is known as 'micromechanics'.

When calculating the mechanical properties of composites it is convenient to start by considering a composite in which all the fibres are aligned in one direction (i.e. a unidirectional composite). This basic 'building block' can then be used to predict the behaviour of continuous fibre multidirectional laminates, as well as short fibre, non-aligned systems.

The essential point about a unidirectional fibre composite is that its stiffness (and strength) are different in different directions. This behaviour contrasts with a metal with a random orientation of grains, or other isotropic material, which has the same elastic properties in all directions.

In a unidirectional composite the fibre distribution implies that the behaviour is essentially isotropic in a cross-section perpendicular to the fibres (Fig. 3.1). In other words, if we were to conduct a mechanical test by applying a stress in the '2' direction or in the '3' direction (both normal to the fibre's longitudinal axis), we would obtain the same elastic properties from each test. We say the material is 'transversely isotropic'. Clearly the properties in the longitudinal ('1') direction are very different from those

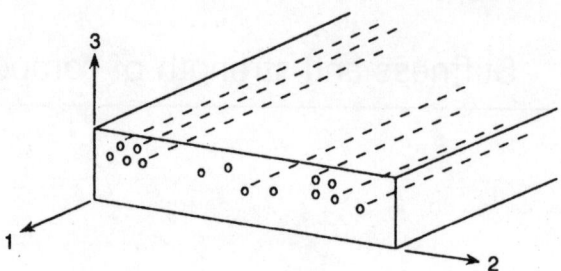

3.1 Orientation of principal material axes.

in the other two directions. We call such a material 'orthotropic'. The elastic properties are symmetric with respect to the chosen (1–2–3) axes, which are usually called the 'principal material axes'.

3.1.2 Basic stress–strain relations

The stress–strain relations for the unidirectional material can readily be found, provided we take account of the fact that the properties are direction-dependent. Considering the composite illustrated in Fig. 3.1, we see that if the directions of the applied stresses coincide with the principal material axes ('specially orthotropic'), the strains in terms of the stresses are given by

$$\varepsilon_1 = \frac{\sigma_1}{E_{11}} - \upsilon_{21}\frac{\sigma_2}{E_{22}}$$

$$\varepsilon_2 = -\upsilon_{12}\frac{\sigma_1}{E_{11}} + \frac{\sigma_2}{E_{22}} \qquad\qquad [3.1]$$

$$\gamma_{12} = \frac{\tau_{12}}{G_{12}}$$

where E_{11} is the elastic modulus in the '1', or longitudinal direction
E_{22} is the elastic modulus in the '2', or transverse direction
G_{12} is the shear modulus in the 1–2 axes
υ_{12} is the 'major' Poisson's ratio
υ_{21} is the 'minor' Poisson's ratio

It should be clear that υ_{12} gives the transverse ('2'- direction) strain caused by a strain applied in the longitudinal ('1') direction; conversely for υ_{21}. Because of the presence of the high stiffness fibres we would intuitively expect υ_{12} to be larger than υ_{21}. This is confirmed by a fundamental law of elasticity which shows that

Table 3.1 Representative elastic properties of unidirectional fibre-reinforced epoxy resins

Material	Fibre volume fraction v_f	E_{11} (GPa)	E_{22} (GPa)	v_{12}	G_{12} (GPa)
CFRP (AS fibre)	0.66	138.0	8.96	0.30	7.10
CFRP (IM6 fibre)	0.65	200.0	11.10	0.32	8.35
GFRP (E-glass fibre)	0.46	39.5	8.22	0.26	4.10
KFRP (Kevlar-49 fibre)	0.60	76.0	5.50	0.33	2.35

$$\frac{v_{12}}{E_{11}} = \frac{v_{21}}{E_{22}}$$

or $\quad v_{12} = v_{21}\dfrac{E_{11}}{E_{22}}$ [3.2]

and as $\quad \dfrac{E_{11}}{E_{22}} > 1, v_{12} > v_{21}$

To get stresses in terms of strains we can rearrange equation [3.1] to give

$$\sigma_1 = \frac{E_{11}\varepsilon_1}{1 - v_{12}v_{21}} + \frac{v_{21}E_{11}\varepsilon_2}{1 - v_{12}v_{21}}$$

$$\sigma_2 = \frac{v_{12}E_{22}\varepsilon_1}{1 - v_{12}v_{21}} + \frac{E_{22}\varepsilon_2}{1 - v_{12}v_{21}}$$ [3.3]

$$\tau_{12} = G_{12}\gamma_{12}$$

Representative, experimentally determined, values of elastic properties are given in Table 3.1 for carbon, glass and Kevlar fibre-reinforced epoxy resins (CFRP, GFRP, KFRP respectively). As we shall see later, it is sometimes possible to calculate elastic constants from the properties of the fibre and matrix using micromechanics.

The important point to note from equations [3.1] and [3.3] is that we need *four* elastic constants to characterise our unidirectional composite; E_{11}, E_{22}, v_{12} (or v_{21}) and G_{12}. Contrast this with an isotropic material where only *two* quantities are needed. For the composite the shear modulus cannot be calculated from E and v, as it can for isotropic materials.

For convenience when dealing with laminates it is helpful to rewrite equations [3.1] and [3.3] in matrix form. So [3.1] becomes

$$\varepsilon_{12} = \mathbf{S}\sigma_{12} \tag{3.4}$$

where $\varepsilon_{12} = \{\varepsilon_1 \ \varepsilon_2 \ \gamma_{12}\}$

and $\sigma_{12} = \{\sigma_1 \ \sigma_2 \ \tau_{12}\}$

Note that { } denotes a column vector written as a row vector and we define the compliance matrix

$$\mathbf{S} = \begin{bmatrix} \dfrac{1}{E_{11}} & -\dfrac{\upsilon_{21}}{E_{22}} & 0 \\[2mm] -\dfrac{\upsilon_{12}}{E_{11}} & \dfrac{1}{E_{22}} & 0 \\[2mm] 0 & 0 & \dfrac{1}{G_{12}} \end{bmatrix}$$

Note that $S_{12} = S_{21}$
and [3.3] becomes

$$\sigma_{12} = \mathbf{Q}\varepsilon_{12} \tag{3.5}$$

where the stiffness matrix is defined by

$$\mathbf{Q} = \begin{bmatrix} \dfrac{E_{11}}{1-\upsilon_{12}\upsilon_{21}} & \dfrac{\upsilon_{21}E_{11}}{1-\upsilon_{12}\upsilon_{21}} & 0 \\[2mm] \dfrac{\upsilon_{12}E_{22}}{1-\upsilon_{12}\upsilon_{21}} & \dfrac{E_{22}}{1-\upsilon_{12}\upsilon_{21}} & 0 \\[2mm] 0 & 0 & G_{12} \end{bmatrix}$$

Note that $Q_{12} = Q_{21}$.

We see that the first column of $\mathbf{S}$ gives the strains caused by a unit value of σ_1, the second column of $\mathbf{Q}$ gives the stresses needed to cause a unit value of ε_2, and so on. It should be clear from the properties of matrices that $\mathbf{Q} = \mathbf{S}^{-1}$.

3.1.3 Off-axis loading of a unidirectional composite

In later sections we shall refer frequently to laminates. As noted in Chapter 2, these plate-like entities are often constructed by assembling layers (or laminae, or plies), usually unidirectional, one on top of another, the direction of the fibres normally being changed from layer to layer. Consequently there will be layers for which the fibres are no longer aligned with the applied stresses (the situation considered in the previous section). We term these 'rotated layers' and say that they are subjected to 'off-axis loading'.

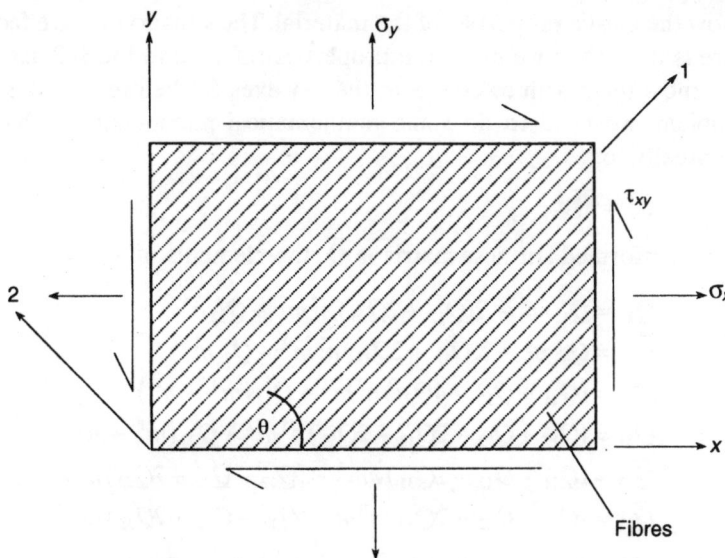

3.2 Unidirectional lamina with principal axes rotated by θ relative to the *x*-*y* axes.

To prepare ourselves for the analysis of laminates, it is useful at this stage to consider in isolation one of these rotated laminae. The situation, which is illustrated in Fig. 3.2, is seen to correspond to the analysis of principal stress for isotropic materials.

We can write, then

$$\sigma_{12} = \mathbf{T}\sigma_{xy} \qquad\qquad [3.6]$$

and $$\bar{\varepsilon}_{12} = \mathbf{T}\bar{\varepsilon}_{xy} \qquad\qquad [3.7]$$

where $\sigma_{12} = \{\sigma_1 \ \sigma_2 \ \tau_{12}\}$

$\sigma_{xy} = \{\sigma_x \ \sigma_y \ \tau_{xy}\}$

$\bar{\varepsilon}_{12} = \{\varepsilon_1 \ \varepsilon_2 \ \tfrac{1}{2}\gamma_{12}\}$

$\bar{\varepsilon}_{xy} = \{\varepsilon_x \ \varepsilon_y \ \tfrac{1}{2}\gamma_{xy}\}$

and the transformation matrix

$$\mathbf{T} = \begin{bmatrix} m^2 & n^2 & 2mn \\ n^2 & m^2 & -2mn \\ -mn & mn & (m^2 - n^2) \end{bmatrix} \qquad\qquad [3.8]$$

with $m = \cos\theta$ and $n = \sin\theta$.

As with any elasticity analysis we wish to determine the strains for a known set of applied stresses (or vice versa). We can do this provided we know the elastic properties of the material. The situation we are faced with here is that while we know the properties referred to the 1–2 axes, we do not know them with reference to the x–y axes. So, before we can solve the problem, we need to do some mathematical manipulation, which leads eventually to:[1]

$$\sigma_{xy} = \overline{Q}\varepsilon_{xy} \qquad\qquad [3.9]$$

The transformed stiffness matrix is $\overline{Q}$, the elements of which are

$$\overline{Q}_{11} = Q_{11}m^4 + 2(Q_{12} + 2Q_{33})n^2m^2 + Q_{22}n^4$$
$$\overline{Q}_{22} = Q_{11}n^4 + 2(Q_{12} + 2Q_{33})n^2m^2 + Q_{22}m^4$$
$$\overline{Q}_{12} = (Q_{11} + Q_{22} - 4Q_{33})n^2m^2 + Q_{12}(m^4 + n^4)$$
$$\overline{Q}_{33} = (Q_{11} + Q_{22} - 2Q_{12} - 2Q_{33})n^2m^2 + Q_{33}(m^4 + n^4)$$
$$\overline{Q}_{13} = (Q_{11} - Q_{12} - 2Q_{33})nm^3 + (Q_{12} - Q_{22} + 2Q_{33})n^3m$$
$$\overline{Q}_{23} = (Q_{11} - Q_{12} - 2Q_{33})n^3m + (Q_{12} - Q_{22} + 2Q_{33})nm^3 \qquad [3.10]$$

We see that knowledge of the orientation (θ) and unidirectional properties (Q) in the principal directions enables us to calculate the stiffness of the rotated lamina. In the general form derived we would call this a 'generally orthotropic lamina'. If conditions are such that $\overline{Q}_{13} = \overline{Q}_{23} = 0$, we have what is called a 'specially orthotropic lamina'. If we require strains in terms of stresses then we invert equation [3.9] to give

$$\varepsilon_{xy} = \overline{Q}^{-1}\sigma_{xy}$$
$$[3.11]$$
or $\qquad \varepsilon_{xy} = \overline{S}\sigma_{xy}$

where $\overline{S}$ is the transformed compliance matrix, the elements of which can be obtained by a similar process to that used for finding the elements of $\overline{Q}$, i.e.

$$\overline{S}_{11} = S_{11}m^4 + (2S_{12} + S_{33})n^2m^2 + S_{22}n^4$$
$$\overline{S}_{22} = S_{11}n^4 + (2S_{12} + S_{33})n^2m^2 + S_{22}m^4$$
$$\overline{S}_{12} = (S_{11} + S_{22} - S_{33})n^2m^2 + S_{12}(m^4 + n^4)$$
$$\overline{S}_{33} = 2(2S_{11} + 2S_{22} - 4S_{12} - S_{33})n^2m^2 + S_{33}(m^4 + n^4)$$
$$\overline{S}_{13} = (2S_{11} - 2S_{12} - S_{33})m^3n + (2S_{12} - 2S_{22} + S_{33})mn^3$$
$$\overline{S}_{23} = (2S_{11} - 2S_{12} - S_{33})n^3m + (2S_{12} - 2S_{22} + S_{33})m^3n \qquad [3.12]$$

For general loading of an off-axis ply we see that in addition to an extension in the direction of the applied stress and a lateral contraction, as we would expect, there is also a shear strain. This phenomenon, which is known as extension-shear coupling, would not be observed with an isotropic material. We see that the shear strain is determined by the $\overline{S}_{13}$ term, when only

σ_x is acting. Had we applied only σ_y, there would have been a shear strain arising from the $\bar{S}_{23}$ term. In other words there will be no extension-shear coupling if $\bar{S}_{13} = \bar{S}_{23} = 0$. This is clearly the case when the fibres of a unidirectional layer are parallel to the stress axes, i.e. $\theta = 0$ or $90°$. It is also possible for $\bar{S}_{13}$ and $\bar{S}_{23}$ to be zero with fibres at intermediate angles; whether, or not, this occurs depends on the relative values of E_{11}, E_{22} and G_{12}.

3.1.4 Stiffness of laminates

Thin sheet constructions, known as laminates, are an important class of composite. They are made by stacking together, usually, unidirectional layers (also called plies or laminae) in predetermined directions and thicknesses to give the desired stiffness and strength properties. Such constructions are frequently encountered. The skins of aeroplane wings and tails, the hull sides and decking of ships, the sides and bottom of water tanks are typical examples. Even cylindrical components, such as filament wound tanks, can be treated as laminates, provided the radius-to-thickness ratio is sufficiently large (say >50). Laminates will be typically between 4 and 40 layers, each ply being around 0.125 mm thick if it is carbon or glass fibre/epoxy prepreg. Typical lay-ups (the arrangement of fibre orientations) are cross-ply, angle ply and quasi-isotropic.

When making a laminate we must decide on the order in which the plies are placed through the thickness (known as the stacking sequence). This has an important influence on the flexural performance of the laminate.

There is an established convention for denoting both the lay-up and stacking sequence of a laminate. Thus, a four-ply cross-ply laminate which has ply fibre orientations in the sequence $0°, 90°, 90°, 0°$ from the upper to the lower surface, would be denoted $(0/90°)_s$. The suffix 's' means that the stacking sequence is symmetric about the mid-thickness of the laminate. Laminates denoted by $(0/45/90°)_s$ and $(45/90/0°)_s$ have the same lay-up but different stacking sequences.

The way in which they are used means that laminated plates may be subjected to both in-plane and transverse (normal to the plate) loading. In other words they will stretch and bend, and both these effects must be taken into account when describing the overall behaviour of the plate. The two effects are separated by considering the total strain to be the superposition of in-plane strains, $\varepsilon°$, constant across the plate thickness, and strains caused by bending, linear across the thickness, as shown in Fig. 3.3. The bending strains can be defined in terms of the plate curvatures. For example, $\varepsilon_x = z\kappa_x$, z being the coordinate normal to the plate measured from the laminate mid-plane, and is positive downwards (as shown in Fig. 3.3).

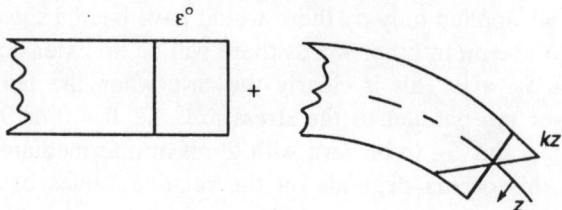

3.3 The two components of laminate strain: ε°, in-plane, constant over the thickness; $\varepsilon_x = zk_x$, bending, linear variation over the thickness.

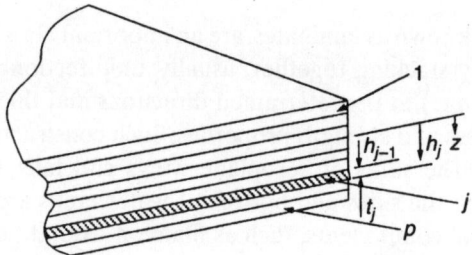

3.4 Definition of plies within a laminate.

So we have

$$\begin{bmatrix} \varepsilon_x \\ \varepsilon_y \\ \gamma_{xy} \end{bmatrix} = \begin{bmatrix} \varepsilon^\circ_x \\ \varepsilon^\circ_y \\ \gamma^\circ_{xy} \end{bmatrix} + z \begin{bmatrix} \kappa_x \\ \kappa_y \\ \kappa_{xy} \end{bmatrix}$$ [3.13]

or $\varepsilon_{xy} = \varepsilon^\circ + z\kappa$ [3.14]

Because all the plies are bonded together in the manufacturing process, we assume that they each have the same in-plane strains and curvatures. So, for any one layer, say the **j**th (Fig. 3.4) we have, using equation [3.9],

$$\sigma_{xy_j} = \overline{Q}_j \varepsilon^\circ + z\overline{Q}_j \kappa$$ [3.15]

$\overline{Q}_j$ being the transformed stiffness matrix for the layer.

Note that the stresses act in the plane of the laminate. These stresses can be converted to equivalent forces (or stress resultants) acting on a unit width of plate. So, for example, from σ_x we get $N_{x_j} = \sigma_{x_j} t_j$, t_j being the thickness of layer j. If we add up the resultants for all the plies the total must be equal to the external force (per unit width) acting on the plate (see Fig. 3.5). Similarly, the equivalent force on a layer will have a moment about the mid-plane. Again using σ_x only to illustrate the process, we have $M_{x_j} = \sigma_{x_j} t_j z_j$, z_j being the distance from the laminate mid-plane to the mid-thickness of the ply. Adding together these moments for all the plies will give the external moment (per unit width) acting on the plate (see Fig. 3.5).

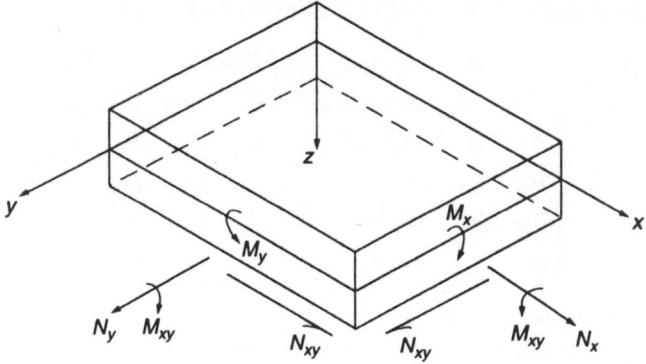

3.5 Loads acting on a laminate.

This procedure allows us to relate the stress resultants to the in-plane strains and curvatures, giving:

$$\mathbf{N} = \mathbf{A}\boldsymbol{\varepsilon}^\circ + \mathbf{B}\boldsymbol{\kappa} \qquad [3.16]$$

and

$$\mathbf{M} = \mathbf{B}\boldsymbol{\varepsilon}^\circ + \mathbf{D}\boldsymbol{\kappa} \qquad [3.17]$$

Equations [3.16] and [3.17] collectively are known as the 'plate constitutive equations', and the associated analysis as 'Classical Laminate Theory'.

The elements of the **A**, **B** and **D** matrices are

$$A_{rs} = \sum_{j=1}^{p} \overline{Q}_{rs_j}[h_j - h_{j-1}] = \sum_{j=1}^{p} \overline{Q}_{rs_j} t_j$$

$$B_{rs} = \frac{1}{2}\sum_{j=1}^{p} \overline{Q}_{rs_j}[h_j^2 - h_{j-1}^2] \qquad [3.18]$$

$$D_{rs} = \frac{1}{3}\sum_{j=1}^{p} \overline{Q}_{rs_j}[h_j^3 - h_{j-1}^3]$$

where $r, s = 1, 3$.

3.1.5 The **A**, **B** and **D** matrices

We have already seen when considering a unidirectional composite that certain terms in the compliance, or stiffness, matrix are associated with coupling between particular deformations and loads. Examination of the plate constitutive equations [3.16] and [3.17] allows us to identify similar couplings for laminates.

Suppose we write [3.16] and [3.17] in expanded form:

$$
\begin{bmatrix} N_x \\ N_y \\ N_{xy} \end{bmatrix} = \begin{bmatrix} A_{11} & A_{12} & A_{13} \\ A_{21} & A_{22} & A_{23} \\ A_{31} & A_{32} & A_{33} \end{bmatrix} \begin{bmatrix} \varepsilon^\circ{}_x \\ \varepsilon^\circ{}_y \\ \varepsilon^\circ{}_{xy} \end{bmatrix} + \begin{bmatrix} B_{11} & B_{12} & B_{13} \\ B_{21} & B_{22} & B_{23} \\ B_{31} & B_{32} & B_{33} \end{bmatrix} \begin{bmatrix} \kappa_x \\ \kappa_y \\ \kappa_{xy} \end{bmatrix}
$$

$$
\begin{bmatrix} M_x \\ M_y \\ M_{xy} \end{bmatrix} = \begin{bmatrix} B_{11} & B_{12} & B_{13} \\ B_{21} & B_{22} & B_{23} \\ B_{31} & B_{32} & B_{33} \end{bmatrix} \begin{bmatrix} \varepsilon^\circ{}_x \\ \varepsilon^\circ{}_y \\ \varepsilon^\circ{}_{xy} \end{bmatrix} + \begin{bmatrix} D_{11} & D_{12} & D_{13} \\ D_{21} & D_{22} & D_{23} \\ D_{31} & D_{32} & D_{33} \end{bmatrix} \begin{bmatrix} \kappa_x \\ \kappa_y \\ \kappa_{xy} \end{bmatrix}
$$

[3.19]

noting, of course, that $A_{21} = A_{12}$, etc.

We can then make the following associations:

- A_{13} and A_{23} relate in-plane direct forces to in-plane shear strain, or in-plane shear force to in-plane direct strains.
- B_{11}, B_{12} and B_{22} relate in-plane direct forces to plate curvatures, or bending moments to in-plane direct strains.
- B_{13} and B_{23} relate in-plane direct forces to plate twisting, or torque to in-plane direct strains.
- B_{33} relates in-plane shear force to plate twisting, or torque to in-plane shear strain.
- D_{13} and D_{23} relate bending moments to plate twisting, or torque to plate curvatures.

In certain circumstances some of the couplings listed above can be undesirable. They can sometimes be eliminated by appropriate construction of the laminate.

If $A_{13} = A_{23} = 0$ there will be no coupling between direct stresses and shear strains (or shear stresses and direct strains). This can be achieved if we have a laminate in which all plies have 0° and/or 90° fibre orientations (a unidirectional or cross-ply laminate), or if the lay-up is *balanced*, i.e. for every layer with a +θ orientation there is an identical lamina with a −θ orientation.

Bending-membrane coupling can be avoided if the B matrix is zero. This is very easily achieved by making the laminate *symmetric* about its midplane. In practice laminates usually have a symmetric stacking sequence.

The phenomenon of bending–twisting coupling is eliminated if $D_{13} = D_{23} = 0$. This is achieved with unidirectional or cross-ply laminates, or with *balanced anti-symmetric* lay-ups, i.e. for every layer at +θ orientation and a given distance above the mid-plane there is a layer with *identical* thickness and properties oriented at −θ and the same distance below the mid-plane. Such a lay-up is *not* symmetric (i.e. $B \ne 0$).

The preference, in practice, for symmetrical laminates means that D_{13} and $D_{23} \ne 0$. However, these terms tend to zero for thick multilayer symmetric laminates.

For some lay-ups *in-plane* behaviour of the laminate is such that it appears to be isotropic (it is then called *quasi-isotropic*). Examples are (0/90/±45°) and (0/±60°) lay-ups. For such laminates it can be shown that

$$A_{11} = A_{22}$$
$$A_{11} - A_{12} = 2A_{33}$$
$$A_{13} = A_{23} = 0$$

3.2   Micromechanics of unidirectional composites

3.2.1 Macromechanics and micromechanics

In the previous section we developed equations that describe the stress–strain behaviour of a lamina and the load–deformation behaviour of a laminate. These equations were based on the elastic properties of the lamina and, as such, ignored the microscopic nature of the material. In other words we took no direct account of the fact that we are dealing with fibre-reinforced materials, we merely acknowledged that they were non-isotropic. We refer to this as macromechanics analysis.

Because the starting point of a significant proportion of composites' manufacture is the combination of fibres and matrix, it would be very helpful if we could predict the behaviour of the composite (laminate) from a knowledge of the properties of the constituents alone. There are many limitations to such micromechanics analyses. However, studying performance on a micro scale is essential if we are to understand fully what controls the strength, toughness, etc., of composites.

3.2.2 Micromechanics models for stiffness

We stated in Section 3.1 that the unidirectional ply forms a useful building block for many studies of composites. Much micromechanics analysis has, therefore, been devoted to this simple system. The most successful application is the prediction of the stiffness parallel to the fibres, i.e. the longitudinal stiffness or modulus. We denoted this as E_{11} in Section 3.1.2.

It can be shown that the longitudinal modulus of the composite is

$$E_{11} = E_f v_f + E_m v_m \qquad [3.20]$$

where subscripts f and m refer to fibres and matrix and v is the volume fraction. A relationship of this form is known as the Rule or Law of Mixtures.

The predictions from equation [3.20] agree well (within 5%) with data from carefully controlled experiments for tensile loading. Predictions are not so good for compressive loading because the experimental results are very sensitive to the design of the equipment and the alignment of the fibres in the specimen.

The corresponding expression for the transverse modulus of the composite is

$$\frac{1}{E_{22}} = \frac{v_f}{E_f} + \frac{v_m}{E_m}$$

[3.21]

It is usual to take E_f as the longitudinal value, if only because the transverse fibre modulus is extremely difficult to determine. This assumption will only be correct for isotropic fibres.

Because of the enormous simplifications made in formulating the above equation [3.21] it does not give very good predictions for E_{22}. Extremely complicated elasticity or finite element analyses are required produce a more accurate model.

The in-plane shear stiffness or modulus, denoted as G_{12} in Section 3.1.2, of a unidirectional composite can be obtained from a similar model to that used for obtaining transverse modulus. The analysis gives the result:

$$\frac{1}{G_{12}} = \frac{v_f}{G_f} + \frac{v_m}{G_m}$$

[3.22]

The model suffers from the same limitations as that used for transverse modulus, very complicated analyses being needed to give a better prediction.

The longitudinal modulus model serves to provide a prediction for the major Poisson's ratio, v_{12}. The resulting expression is

$$v_{12} = v_f v_f + v_m v_m$$

[3.23]

3.2.3 Micromechanics models for strength

It is far more difficult to obtain a prediction for strength than for stiffness. This is due to several factors: the random nature of failure and hence the need to employ statistical methods; the number of failure modes that can cause the composite's failure (fibre, matrix or interface failure); the very local nature of failure initiation and the influence of the associated stress field, itself determined by the details of fibre packing. As for stiffness, methods for predicting longitudinal performance are better than those for transverse and shear performance.

For most PMCs, a reasonable description of tensile strength capability is given by $\hat{\sigma}_{1T} = \hat{\sigma}_f v_f$, but it should be recognised that the onset of cracking (of fibres or matrix) may be a more useful design criterion. Also, fibre strength is not a unique value; it varies from one fibre to another and depends on the length over which it is measured. Thus, the choice of $\hat{\sigma}_f$ is far from obvious and it is necessary to use a statistical approach. The complex nature of tensile failure, involving fibre, matrix or interface failure, is also seen in compression, but with the added possibilities of fibre buck-ling and matrix shear deformation.

Recent work which accounts for initial fibre waviness and the matrix shear yield stress gives reasonably accurate predictions:

$$\hat{\sigma}_{1C} = G_t(\gamma^*)$$

i.e. the shear modulus of the composite at instability.

Failure strains in compression can be very close to those seen in tension; however, results are very dependent on the test method.

There is no simple relation for predicting the transverse tensile strength ($\hat{\sigma}_{2_T}$), transverse compressive strength ($\hat{\sigma}_{2C}$) or in-plane shear strength ($\hat{\tau}_{12}$). Because of the complexities, no serious attempts have been made to model these modes of failure.

Note that there is another form of shear failure, namely *interlaminar shear failure*. This mode of failure is seen when flexural or through-thickness loading occurs (see Section 3.4). It is, then, relevant to *delamination*, i.e. splitting between plies.

3.2.4 Thermal and moisture effects

As already mentioned, carbon and aramid fibres have a very small, or even slightly negative, coefficient of longitudinal thermal expansion (α_f). One consequence of this is that residual stresses are set up in a unidirectional composite as it cools from the curing temperature. In the transverse direction these stresses can be a significant fraction of the failure stress of the matrix. Hence, any calculation which attempts to predict failure should include these thermal effects.

A second consequence of the fibre and matrix having different expansion coefficients is that the composite has different coefficients in the longitudinal and transverse directions. Based on the assumption that the Poisson's ratios $\upsilon_f \approx \upsilon_m$ the following simple expressions have been derived:

$$\alpha_1 = \frac{1}{E_{11}}(\alpha_f E_f v_f + \alpha_m E_m v_m) \tag{3.24}$$

$$\alpha_2 = (1 + \upsilon_f)\alpha_f v_f + (1 + \upsilon_m)\alpha_m v_m - \alpha_1 \upsilon_{12} \tag{3.25}$$

The anisotropy of thermal expansion ($\alpha_1 \neq \alpha_2$) will cause residual thermal stresses in laminates, in addition to those in each ply mentioned above.

Resin matrices will absorb moisture and therefore swell during normal operating conditions. Some fibres, such as Kevlar, also absorb water. The water up-take in a resin matrix is usually by a process of Fickian diffusion and the swelling can be characterised by an absorption coefficient (β), which is directly equivalent to the thermal expansion coefficient.

For a unidirectional PMC, reinforced with carbon or glass fibres, the swelling coefficient in the fibre direction (β_1) may be taken as zero. The transverse coefficient (β_2) depends on the expansion of the matrix, β_m, and may be taken as:

$$\beta_2 = \frac{\rho_c}{\rho_m}(1 + \upsilon_m)\beta_m \qquad [3.26]$$

where the suffices m and c refer to matrix and composite respectively. The composite's density, ρ_c, was defined in equation [2.2].

Laminate analysis should be modified to take account of thermal and moisture effects. This is necessary so that a better estimate of strength may be obtained. The appropriate theory can be found in many texts, e.g. Agarwal and Broutman.[1]

3.3   Strength of unidirectional composites and laminates

3.3.1   Introduction

In Section 3.1 we established the stress–strain relationships for an individual lamina, or ply, and for a laminate. We can use the constitutive equations to calculate the stresses in each ply when we know the values of the loads acting on the laminate. By comparing these stresses with a corresponding limiting value we can decide whether or not the laminate will fail when subjected to the service loads.

There are several ways to define failure. The obvious one is when we have complete separation, or fracture; clearly, then, the component can no longer support the loads acting on it. However, a more general definition would be 'when the component can no longer fulfil the function for which it was designed'.

Such a definition includes total fracture but could also include excessive deflection as seen when a laminate buckles (basically a stiffness rather than a strength limit), or even just matrix cracking. The latter could constitute failure for a container because any contents would be able to leak through the matrix cracks in the container's walls.

As for isotropic materials, a failure criterion can be used to predict failure. A large number of such criteria exist, no one criterion being universally satisfactory. We shall start by considering a single ply before moving on to discuss laminates.

We saw in the last section that there are five basic modes of failure of such a ply: longitudinal tensile or compressive, transverse tensile or compressive, or shear. Each of these modes would involve detailed failure

Table 3.2 Typical strengths of unidirectional PMC laminates ($v_f = 0.5$) (values in MPa)

Material	Longitudinal tension	Longitudinal compression	Transverse tension	Transverse compression	Shear
Glass/polyester	650–750	600–900	20–25	90–120	45–60
Carbon/epoxy	850–1100	700–900	35–40	130–190	60–75
Kevlar/epoxy	1100–1250	240–290	20–30	110–140	40–60

mechanisms associated with fibre, matrix or interface failure. Some typical strengths of PMCs are shown in Table 3.2.

We can regard the strengths in the principal material axes (parallel and transverse to the fibres) as the fundamental parameters defining failure. When a ply is loaded at an angle to the fibres, as it is when part of a multi-directional laminate, we have to determine the stresses in the principal directions and compare them with the fundamental values.

3.3.2 Strength of a lamina

Strength can be determined by the application of failure criteria, which are usually grouped into three classes: limit criteria, the simplest; interactive criteria, which attempt to allow for the interaction of multi-axial stresses; hybrid criteria, which combine selected aspects of limit and interactive methods. Here we shall only discuss criteria that fit into the first two classes.

3.3.2.1 Limit criteria

There are two limit methods, the maximum stress and the maximum strain criteria. The maximum stress criterion consists of five subcriteria, or limits, one corresponding to the strength in each of the five fundamental failure modes. If any one of these limits is exceeded, by the corresponding stress expressed in the principal material axes, the material is deemed to have failed.

In mathematical terms we say that failure has occurred if:

$$\sigma_1 \geq \hat{\sigma}_{1T} \text{ or } \sigma_1 \leq \hat{\sigma}_{1C} \text{ or } \sigma_2 \geq \hat{\sigma}_{2T} \text{ or } \sigma_2 \leq \hat{\sigma}_{2C} \text{ or } \tau_{12} \geq \hat{\tau}_{12} \qquad [3.27]$$

(recalling that a compressive stress is taken as negative so, for example, failure would occur if $\sigma_2 = -200\,\mathrm{MPa}$ and $\hat{\sigma}_{2C} = -150\,\mathrm{MPa}$).

The maximum strain criterion merely substitutes strain for stress in the five subcriteria. We now say that failure has occurred if:

$$\varepsilon_1 \geq \hat{\varepsilon}_{1T} \text{ or } \varepsilon_1 \leq \hat{\varepsilon}_{1C} \text{ or } \varepsilon_2 \geq \hat{\varepsilon}_{2T} \text{ or } \varepsilon_2 \leq \hat{\varepsilon}_{2C} \text{ or } \gamma_{12} \geq \hat{\gamma}_{12} \qquad [3.28]$$

As when calculating stiffness, it is important that we can deal with the situation in which the fibres are not aligned with the applied stresses. We would use equations [3.6] to obtain the stresses in the principal material directions.

Although simple to use, limit criteria do not agree well with experimental data unless the fibre angle is close to 0° or 90°. This is because at intermediate angles there will be a stress field in which both σ_1 and σ_2 can be significant. These stresses will interact and affect the failure load, a situation which is not represented in a criterion where a mode of failure is assumed not to be influenced by the presence of other stresses.

Also, the maximum stress and maximum strain criteria will give different predictions when the stress–strain relation is nonlinear. This will certainly be the case for shear deformations and hence an assumption of linearity is seen to be invalid. In such cases the maximum strain criterion generally gives better agreement with experiment than does the maximum stress criterion.

3.3.2.2 Interactive criteria

Interactive criteria, as the name suggests, are formulated in such a way that they take account of stress interactions. The objective of this approach is to allow for the fact that when a multi-axial stress state exists in the material failure loads may well differ from those when only a uni-axial stress is acting.

There are many such criteria, of varying complexity, their success in predicting failure often being confined to one fibre/resin combination subjected to a well defined set of stresses (e.g. a tube under internal pressure). The Tsai–Hill criterion which has proven to be successful in a wide variety of circumstances is the only method that will be discussed here. There is currently much debate about the merits of the various criteria.[2]

The Tsai–Hill criterion was developed from Hill's anisotropic failure criterion which, in turn, can be traced back to the von Mises yield criterion for metals. In its most general form the Tsai–Hill criterion defines failure as:

$$\left(\frac{\sigma_1}{\hat{\sigma}_1}\right) - \frac{\sigma_1\sigma_2}{\hat{\sigma}_1^{\,2}} + \left(\frac{\sigma_2}{\hat{\sigma}_2}\right)^2 + \left(\frac{\tau_{12}}{\hat{\tau}_{12}}\right)^2 \geq 1 \qquad [3.29]$$

Because it is usually small compared with the others, the second term $(\sigma_1\sigma_2/\hat{\sigma}_1^{\,2})$ is often neglected. The modified from of the criterion is then:

$$\left(\frac{\sigma_1}{\hat{\sigma}_1}\right)^2 + \left(\frac{\sigma_2}{\hat{\sigma}_2}\right)^2 + \left(\frac{\tau_{12}}{\hat{\tau}_{12}}\right)^2 \geq 1 \qquad [3.29a]$$

The values of strength used in equation [3.29] or [3.29a] are chosen to correspond to the nature of σ_1 and σ_2. So if σ_1 is tensile $\hat{\sigma}_{1T}$ is used, if σ_2 is compressive $\hat{\sigma}_{2C}$ would be used, and so on.

It should be noted that only one criterion has to be satisfied, as opposed to the five subcriteria of the limit methods. Thus, only one value is obtained for the failure stress. Another point to bear in mind is that the mode of failure is not indicated by the method, unlike with the limit criteria. This latter issue has an influence on how we predict the failure of laminates, as we shall see later. For a unidirectional composite subjected to uni-axial stress parallel to a principal direction the Tsai–Hill and maximum stress criteria will give the same failure stress. Interactive criteria usually give better predictions of strength for intermediate fibre angles.

3.3.3 Strength of a laminate

3.3.3.1 Initial failure

Suppose we take a cross-ply laminate (0/90° lay-up) and apply an increasing load in the direction of the 0° fibres. At a relatively low load cracks will be seen in the matrix parallel to the fibres in the 90° plies. These cracks will increase in density until a saturation state is reached. At this point the 90° plies contribute virtually no stiffness to the laminate in the 0° direction, a fact that is shown by the change in slope of the load-extension curve for such a laminate. The commencement of transverse ply cracking is known as 'initial failure' or 'first ply failure'.

It is possible to predict initial failure of laminates by combining Classical Laminate Theory with a failure criterion. Clearly, the choice of criterion is crucial and, as already stated, there are many available, each one often being relevant only to a very specific situation (loading and geometry).

We start with the plate constitutive equations (equations [3.16] and [3.17]), i.e.

$$\begin{bmatrix} \mathbf{N} \\ \mathbf{M} \end{bmatrix} = \begin{bmatrix} \mathbf{A} & \mathbf{B} \\ \mathbf{B} & \mathbf{D} \end{bmatrix} \begin{bmatrix} \varepsilon^\circ \\ \kappa \end{bmatrix} \quad\quad\quad [3.30]$$

Solution of equations [3.30] will give the plate mid-plane strains (ε°) and plate curvatures (κ), for a known set of forces $\mathbf{N}$ and moments $\mathbf{M}$.

$$\begin{bmatrix} \varepsilon^\circ \\ \kappa \end{bmatrix} = \begin{bmatrix} \mathbf{A} & \mathbf{B} \\ \mathbf{B} & \mathbf{D} \end{bmatrix}^{-1} \begin{bmatrix} \mathbf{N} \\ \mathbf{M} \end{bmatrix} \quad\quad\quad [3.31]$$

The plate strains are expressed in the, global, plate $\mathbf{x}$–$\mathbf{y}$ axes (see Fig. 3.5). Thus the strains in each ply can be found from the transformations of equation [3.7], and the ply stresses are obtained from the stiffness matrix (equation [3.5]). By applying, on a ply-by-ply basis, a selected failure criterion, the occurrence of failure can be determined.

3.3.3.2   *Final failure and strength*

The final, or ultimate, failure load of an angle ply laminate is often coincident with, or only slightly higher than, the load to cause initial failure. This is not necessarily the case for other lay-ups and final failure can be at a considerably higher load than that to cause first ply failure.

It is clear that once a ply has sustained failure its stiffness in certain directions will have been reduced. However, unless the damaged ply has completely delaminated from the rest of the laminate it will still contribute to the overall stiffness of the plate. The magnitude of this contribution depends on the amount of damage, the fibre/matrix combination and the nature of the loading on the ply.

In general an iterative method is adopted, successively applying the approach described in Section 3.3.3.1 until final failure has occurred. At each step the **A**, **B** and **D** matrices would need to be recalculated to allow for the development of damage.[1]

3.4   Application to structures

3.4.1   Limitations of Classical Laminate Theory

Classical Laminate Theory (CLT) (see Section 3.1.4) applies to an element of plate over which forces and moments are assumed constant, and in which through-thickness shear strains are ignored. Thus, only in-plane direct and shear stresses are considered. CLT is equivalent to simple beam theory which only considers a pure moment loading (i.e. constant along the beam), which results in only a direct stress (acting parallel to the beam's longitudinal axis). Again shear strains (through the beam depth) are ignored.

In practice, of course, beams are usually subjected to bending moments which vary along the length, the resultant direct stresses causing shear stresses, which can be calculated from equations of equilibrium. Even in these circumstances shear strains can usually be ignored; they only have a significant effect on beam deflections when the beam is very short (small span-to-depth ratio). For a laminated construction such shears are referred to as interlaminar shear stresses; they can give rise to interlaminar failure, or delamination. A three-point flexure test on a short beam (span-to-depth = 5) is a common way of determining the interlaminar shear strength of unidirectional composites.

Plate equations can be altered to include varying moments and through-thickness shear strains. The plate constitutive equations would be modified

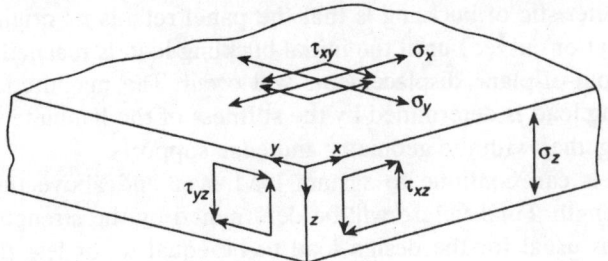

3.6 Stresses at laminate edge in the thickness (*z*) direction.

to account for the through-thickness shear resultants, by including two additional terms in the **A** matrix[3] (see Section 5.5).

The inclusion of shear strains would be appropriate for thick isotropic plates (equivalent to short beams), and also for thick composite plates. The latter, of course, will generally have a low through-thickness shear modulus (compared with in-plane extensional moduli), thus making the inclusion of such shear strains generally advisable for typical composites, even if of moderate thickness.

3.4.2 Edges

The limitations of CLT mentioned above (constant in-plane stresses) implies that the plate is infinitely long and wide. In other words the theory ignores edges. In many real situations, of course, laminates will have edges; for example a plate of finite width (encountered in mechanical testing), a plate containing a hole (bolts, rivets) or opening (say for access), a skin/stringer joint, a skin/spar junction, etc. The stresses acting at an edge are shown in Fig. 3.6.

A simple example is that of the free edge of a cross-ply laminate. Because of the mismatch in ply Poisson's ratio (90/0°)$_s$ is preferred to (0/90°)$_s$ because the latter produces tensile through-thickness stresses.

A problem in analysing these situations is that the high stresses occur within one or two plate thicknesses from the edge. Also, solutions can indicate singularities. Finite element models of this situation are discussed in Chapter 7.

3.4.3 Buckling

In addition to the forms of failure already described, composite laminates, especially if thin, can fail by buckling when subjected to compressive or shear loading.

The characteristic of buckling is that the panel retains its original configuration (flat or curved) until the initial buckling load is reached. At this point large out-of-plane displacements will occur. The magnitude of the corresponding load is determined by the stiffness of the laminate (not the strength), together with the geometry and edge supports.

Most panels can continue to sustain load over and above the initial buckling strength. Final failure will be determined by the strength of the laminate. It is usual for the design load to be equal to, or less than, the initial buckling load.

An extensive set of information on panel buckling can be obtained from: ESDU International plc, 27 Corsham Street, London N1 6UA, UK. An excellent comprehensive reference is that by Leissa.[4]

Only in a few situations is it possible to use simple formulae to calculate initial buckling loads. More often it is necessary to use graphical, or other means. In this section only a few cases will be presented, merely to indicate the general approach.

For convenience it is common to cast the equations into the same form as those used for isotropic plates, i.e. critical stress

$$\sigma_b = KE(t/b)^2$$

where t = plate thickness, b = plate width, E = Young's modulus and K is a buckling coefficient related to panel dimensions and edge supports. Most solutions assume that the panel buckles into a sinusoidal shape with m half waves along the length (x-direction) and n half waves across the width (y-direction). Normally $n = 1$. K is dependent also on m and n.

So, for example, for a rectangular orthotropic plate ($a \times b$) with all edges simply supported, subjected to a uniform uniaxial compressive stress (σ_x), the buckling coefficient is:

$$K_x = \frac{\sigma_x t b^2}{D_{22}} = \frac{N_x b^2}{D_{22}} \qquad [3.32]$$

and $\qquad \dfrac{K_x}{\pi^2} = \dfrac{D_{11}}{D_{22}}\left(\dfrac{b}{a}\right)^2 m^2 + 2\left(\dfrac{D_{12}}{D_{22}} + \dfrac{2D_{33}}{D_{22}}\right) + \left(\dfrac{a}{b}\right)^2 \dfrac{1}{m^2} \qquad [3.33]$

Note the dependence of σ_x on the plate's flexural coefficients, D_{11}, etc. Clearly, an analysis that ignored stacking sequence would not be expected to deliver accurate predictions.

3.5   Summary

In this chapter we have seen now the conventional stress–strain equations in two dimensions are modified to account for the directional nature of composites. From this macromechanics starting point equations describing the behaviour of laminated plates were derived. As pointed out at the end of the chapter, these equations have certain limitations and cannot be used to represent behaviour at boundaries (plate edges or holes).

The equations of micromechanics were shown to have limitations and these should be borne in mind when using software that takes fibre and matrix properties as input data.

3.6　References

1 Agarwal B D & Broutman L J, *Analysis and Performance of Fiber Composites* (2nd Edition), New York, Wiley, 1990.
2 'Failure Criteria in Fibre-Reinforced Polymer Composites', Special Issue, *Composites Science & Technology*, 1998 **58**(7).
3 Vinson J R & Sierakowski R L, *The Behaviour of Structures Composed of Composite Materials*, Dordrecht, Martinus Nijhoff, 1986.
4 Leissa A W, *Buckling of Laminated Composite Plates and Shallow Panels*, US Air Force Flight Dynamics Laboratory, Technical Report AFWAL-TR-85-3069, June 1985.

Part II
Fundamentals of finite element analysis

G. A. O. DAVIES

In this part of the book the basis of finite element analysis is reviewed. Starting from the underlying theoretical issues, the method is developed to show how elements are derived, a model is generated and results are processed.

Fundamentals of finite element analysis

4.1 Basic theoretical foundations

This chapter is not intended to probe deeply into the theory of finite elements. Readers of this book may be familiar with the FE method and interested only in applying it to composite materials and structures, in which case they can skip this chapter. For others, with a passing knowledge only, we shall condense the theory to the basics. It is useful to strip some of the mystery from the theory and to draw attention to what is important and what is unimportant – particularly what is exact and what is approximate. There are dangers in creating an unsound finite element model and not to recognise poor answers.

However, it is counter-productive to simplify the method too much and encourage users of commercial codes to treat the FE package as a 'black box'. User manuals encourage analysts to think of finite elements as pieces in a jigsaw, the whole of which is the complete structure. Thus the package merely has to insert the jigsaw pieces into the correct place and they will be held there by nodal displacements and nodal forces at selected nodes. In the complete structure, when loaded, the displacements at nodes will ensure continuity and the nodal forces developed as the jigsaw piece strains will be in equilibrium with themselves and with the applied forces – somehow. There are two dangers to this approach.

Firstly, the user has no idea what is exact and what is approximate, and in the latter case whether the errors are significant. Secondly, what are these nodal forces and do they not cause a local stress concentration in a stress field that should otherwise be smooth and changing gradually? The answer is that these nodal forces are quite fictitious in the usual FE method, even though they are made to equilibrate each other and the applied forces. (Although the applied forces may have to be converted to a compatible but fictitious form also!)

Having said we shall strip away some of the mystery, the treatment will still be brief. For readers who wish for more we shall use the methods and

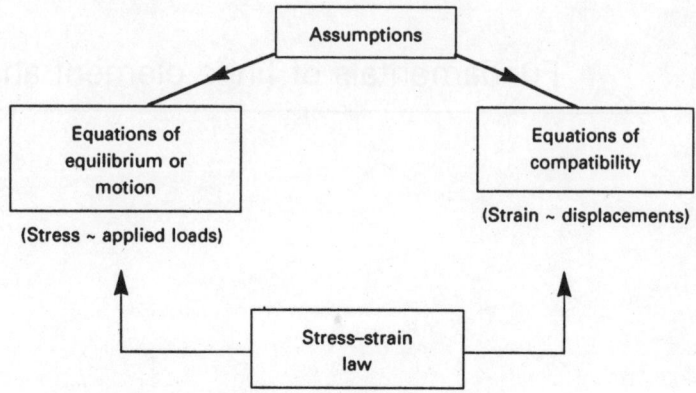

4.1 The four conditions for *all* structural analysis.

notation of *A Finite Element Primer*.[1] We make a beginning by going to the very basics of structural analysis, starting with a trivial example and then generalising to a completely arbitrary structure. Basically all structural analysis can be summarised by four separate (but linked) conditions to be modelled and satisfied, as illustrated in Fig. 4.1.

The basic *assumptions* are usually directed at simplifying the *type of structure or component*. Thus is the component a rod/bar, beam, plate, shell or 3D? This stage leads to the justification – fundamental to the FE method – of assuming the nature of the strain or displacement field in the element. A pin-ended bar in a framework may have only one axial strain component. A slender beam or thin plate may have a linear strain distribution through the thickness. A 3D continuum may not be accessible to any assumed field. Thus this stage is driven by geometry, but we delay the introduction to finite elements proper and consider the other three stages.

Satisfying *equilibrium* results in equations that equilibrate the applied forces to the structure's resisting stresses. These may be surface forces or internal (body) forces. If the structure is in motion then the body forces can be extended to include the 'inertia forces' and of course the stresses, displacements and so on are then all functions of time. In many cases (probably most) the strains are small and it is possible to write down the equations of equilibrium for the body ignoring its small changes in shape. We shall do this initially, and leave to later chapters the nonlinear behaviour due to material failure, or to buckling when changes in geometry cannot be ignored.

If the equations of equilibrium are sufficient to enable us to solve completely the stresses in the structure, the structure is said to be *statically determinate*. In practice very few structures are made to be statically determinate. Such structures have one distinct disadvantage in that if a single component fails, the complete structure becomes a mechanism, i.e. it falls down. A statically determinate structure is very much not 'fail-safe'. There is at least one case where statical determinacy is an advantage, and that is if thermal heating is significant. A statically determinate structure's stresses depend only on the applied loads and are unaffected by temperature gradients. (The ribs inside Concorde are made this way, since the fuel in the wing is intended to cool down the very hot outer surfaces.)

If a structure is *statically indeterminate* or *redundant* then the next two conditions and equations have to be brought in.

Compatibility arises from stating that the strains are 'compatible' with the displacements and can be derived from them if the displacements vary continuously. We show that it simply means that a geometrical argument can be used to express strains in terms of displacements once we have defined what strain means: like the strain in a bar is its fractional change in length. The equations of compatibility have nothing to do with the equations of equilibrium nor with the stresses directly. Thus if the body is heated and strains and stresses ensue, the equations of compatibility are not affected.

The *stress–strain law* is the relationship between stress and strain and it is based on experimental evidence. In many cases stress is proportional to strain. In other cases strain can be caused by heating (or freezing), or by moisture absorption or by electromagnetic-chemical effects. It is important to note that the stress–strain law has nothing to do with either equilibrium or compatibility.

To illustrate these separate criteria consider the very simple example shown in Fig. 4.2 of three pin-ended bars subjected to loads R_1 and R_2 and having joint displacements r_1 and r_2. Let the central bar also be heated by T degrees.

The *assumption* is that all bars are pin-ended, and hence only resist axial tensions T, and stretch by Δ.

Equilibrium:

$$T_{14}\sin\theta - T_{34}\sin\theta = R_1$$
$$T_{14}\cos\theta + T_{24} + T_{34}\cos\theta = R_2 \tag{4.1}$$

Compatibility: if the displacements are small, then the displacement components at the ends of a bar, *at right angles to the bar*, will not stretch it. Only the components *in line* with the bar will produce strains. Thus strain

$$\varepsilon_{14} = \frac{\Delta_{14}}{L}$$

where $\Delta_{14} = r_1\sin\theta + r_2\cos\theta$

$$\Delta_{24} = \frac{r_2}{L\cos\theta} \tag{4.2}$$

and $\Delta_{34} = -r_1\sin\theta + r_2\cos\theta$

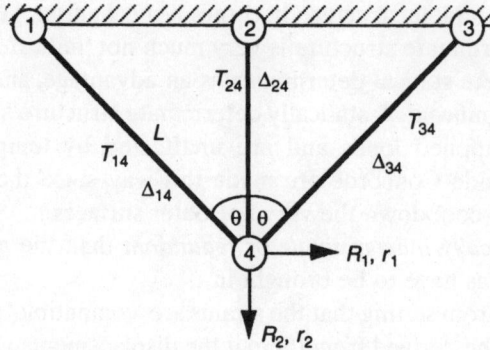

4.2 Simple (redundant) pin-jointed frame example.

Stress–strain:

For outer bars $\varepsilon = \dfrac{\sigma}{E}$

For centre bar $\varepsilon = \dfrac{\sigma}{E} + \alpha T$ [4.3]

where E is the extensional (Young's) modulus and α the linear coefficient of expansion.

A *displacement* solution is now straightforward. If all bars have the same cross-section area A, the stress–strain law becomes

$$\frac{\Delta_{14}}{L} = \frac{T_{14}}{AE}; \quad \frac{\Delta_{24}}{L\cos\theta} = \frac{T_{24}}{AE} + \alpha T; \quad \frac{\Delta_{34}}{L} = \frac{T_{34}}{AE} \qquad [4.4]$$

and using compatibility (equation [4.2]):

$$T_{14} = \frac{AE}{L}[r_1 \sin\theta + r_2 \cos\theta]$$

$$T_{24} = \frac{AEr_2}{L\cos\theta} - AE\alpha T \qquad [4.5]$$

$$T_{34} = \frac{AE}{L}[-r_1 \sin\theta + r_2 \cos\theta]$$

Now we put these forces into the two equations of equilibrium [4.1] and solve for r_1 and r_2:

$$\frac{2AEr_1}{L}\sin^2\theta = R_1$$

$$\frac{AE}{L}(2\cos^2\theta + \sec\theta)r_2 = R_2 - AE\alpha T \qquad [4.6]$$

Notice that the final set of equations (two) were just those needed to solve for the two displacement unknowns. This would have been true if there had been only two bars (statically determinate) or three or more (progressively more and more redundant). This is the attraction of the displacement method; we do not have to worry about the degree of redundancy.

The only difference between this example and the conventional FE solution (apart from the small number of unknowns r_1 and r_2) is that the 'elements' are exact. No assumptions were made about the nature of the displacements since uniform pin-ended bars have a constant strain.

4.2   Principle of virtual displacements

The next step is to use virtual work arguments as a substitute for equilibrium conditions (the advantages will become apparent later). If a force (or a stress) moves through a displacement (or a strain) then the virtual work is defined as the product of force × displacement. This avoids having to define the force–displacement (or stress–strain) law which has nothing to do with equilibrium. The force–displacement relationship may be nonlinear, as shown in Fig. 4.3 for example.

We now go further and imagine *virtual displacements* which are hypothetical and not related to the real forces or displacements. We use a 'bar' to single them out; $\bar{r}_1$, $\bar{r}_2$, $\bar{\Delta}$, etc.

The Principle of Virtual Displacements (PVD) simply equates internal virtual work to external virtual work, thus in the pin-jointed framework example:

$$T_{14}\bar{\Delta}_{14} + T_{24}\bar{\Delta}_{24} + T_{34}\bar{\Delta}_{34} = R_1\bar{r}_1 + R_2\bar{r}_2 \qquad [4.7]$$

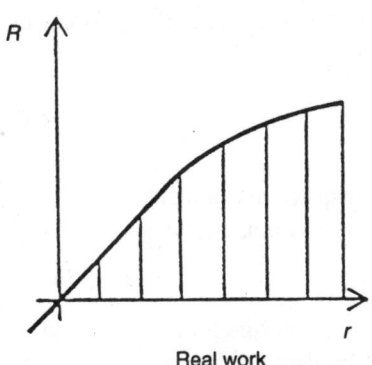
Real work

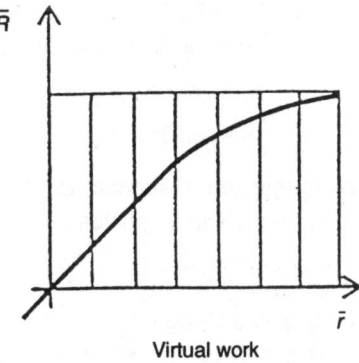
Virtual work

4.3 Nonlinear force/displacement behaviour.

This rather important equation gives us no information at all (!) until we use the compatibility conditions. We satisfy compatibility directly and insert the values for $\bar{\Delta}$ in terms of $\bar{r}$ into the above, and rearrange

$$(T_{14}\sin\theta - T_{34}\sin\theta - R_1)\bar{r}_1 + (T_{14}\cos\theta - T_{24} + T_{34}\cos\theta - R_2)\bar{r}_2 = 0 \qquad [4.8]$$

We now capitalise on the virtual nature of $\bar{r}_1$ and $\bar{r}_2$ and say for example that $\bar{r}_1$ could be zero, in which case the individual expressions in the parenthesis above must *separately vanish*. These are clearly the two equations of equilibrium as before (equation [4.1]) so the PVD is enforcing them indirectly. Three important points to note are:

- Only because $\bar{r}_1$ and $\bar{r}_2$ are independent do we get both equilibrium equations.
- If a framework had n displacements we would get n equations from the n virtual displacements to solve from.
- From the displacement in the (say) horizontal direction, $\bar{r}_1$, we get the horizontal equation of equilibrium, resolving T_{14} and T_{34} in terms of R_1.

Having fixed the ideas we can now move to general structures with arbitrary shapes.

4.3   An arbitrary structure

The leap from a simple three-bar truss to a completely general structure is a big one. We shall need more sophisticated tools to describe the structure and its loads, stresses, and displacements. It turns out that matrix notation is a great aid in defining both stress fields and displacement fields and, moreover, matrices are an ideal vehicle for reading into, or extracting from, digital computer memories. In fact, in the early days of the 1950s, finite element methods were *originally* known as 'matrix methods'.

Figure 4.4 indicates an arbitrary 3D solid having volume V and surface S (local normal 'n'), subjected to surface forces p_s per unit area and internal (volume) body forces p_v per unit volume. We list these vector forces as having Cartesian components:

$$\begin{aligned} \mathbf{p}_v^t &= [p_{vx}\ p_{vy}\ p_{vz}] \\ \mathbf{p}_s^t &= [p_{sx}\ p_{sy}\ p_{sz}] \end{aligned} \qquad [4.9]$$

Note that the 't' superscript denotes the transpose of a matrix.

Similarly the displacements everywhere we write as the vector

$$\mathbf{u}^t = [u\ v\ w] \qquad [4.10]$$

Notice that these components p, u, etc., may all be functions of x, y and z.

The internal stresses and strains can also be listed as column matrices in terms of the three direct and three shear components:

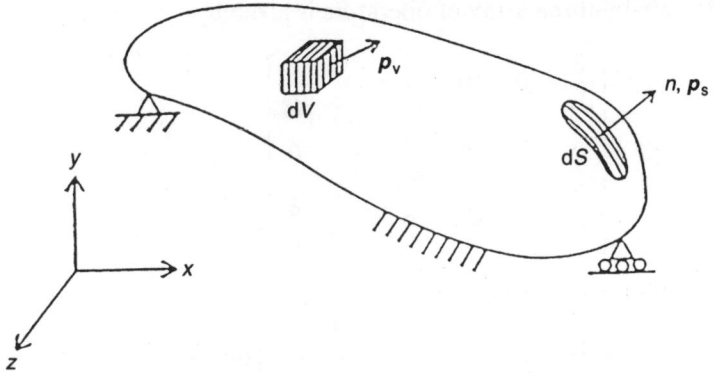

4.4 A general continuum.

$$\sigma^t = [\sigma_{xx}\sigma_{yy}\sigma_{zz}\sigma_{xy}\sigma_{yz}\sigma_{zx}]$$

$$\varepsilon^t = [\varepsilon_{xx}\varepsilon_{yy}\varepsilon_{zz}\varepsilon_{xy}\varepsilon_{yz}\varepsilon_{zx}]$$

[4.11]

The reader can consult any standard text for definitions of stress notation. Suffice it to say that here the first suffix refers to the direction, and the second suffix the plane on which the stress component acts. Thus σ_{xx} is a direct stress (tensile positive, compressive negative) but σ_{xy} is a shear stress. The strain components simply *correspond* to the stress components.

We now follow the procedure as before. The external virtual work done is

$$\int_S (p_{sx}\bar{u} + p_{sy}\bar{v} + p_{sz}\bar{w})dS + \int_V (p_{vx}\bar{u} + p_{vy}\bar{v} + p_{vz}\bar{w})dV$$

$$= \int_S \mathbf{p}_s^t \bar{\mathbf{u}}dS + \int_V \mathbf{p}_v^t \bar{\mathbf{u}}dV$$

[4.12]

It can be shown that the internal virtual work done by the stresses over the strains is

$$\int_V (\sigma_{xx}\bar{\varepsilon}_{xx} + \sigma_{yy}\bar{\varepsilon}_{yy} + \sigma_{zz}\bar{\varepsilon}_{zz} + \sigma_{xy}\bar{\varepsilon}_{xy} + \sigma_{yz}\bar{\varepsilon}_{yz} + \sigma_{zx}\bar{\varepsilon}_{zx})dV = \int_V \sigma^t \bar{\varepsilon}\,dV$$

[4.13]

The next step, as in the framework example, is to equate the two expressions for work and enforce compatibility. The general equations relating ε and $\mathbf{u}$ are readily shown to be, using geometrical definitions of strain in terms of the displacement at a point,

$$\varepsilon_{xx} = \frac{\partial u}{\partial x}, \quad \varepsilon_{xy} = \frac{\partial u}{\partial y} + \frac{\partial v}{\partial x}, \text{etc.}$$

[4.14]

or simply using matrix shorthand $\varepsilon = \partial\mathbf{u}.$

[4.15]

The six-by-three array of operators is given by

$$\partial^t = \begin{bmatrix} \dfrac{\partial}{\partial x} & 0 & 0 & \dfrac{\partial}{\partial y} & 0 & \dfrac{\partial}{\partial z} \\[2ex] 0 & \dfrac{\partial}{\partial y} & 0 & \dfrac{\partial}{\partial x} & \dfrac{\partial}{\partial z} & 0 \\[2ex] 0 & 0 & \dfrac{\partial}{\partial z} & 0 & \dfrac{\partial}{\partial y} & \dfrac{\partial}{\partial x} \end{bmatrix} \qquad [4.16]$$

The PVD then becomes

$$\int_V \sigma^t \bar{\varepsilon}\, dV = \int_V \sigma^t \partial\, \bar{u}\, dV = \int_V p_v^t \bar{u}\, dV + \int_S p_s^t \bar{u}\, dS \qquad [4.17]$$

This most important equation is the basis behind all FE derivations. Later we show what it leads to. To show what we are actually doing when we use it, it is necessary to transform the equation by using a mathematical identity derived from Gauss's Theorem:

$$\int_V \sigma^t \partial\, \bar{u}\, dV = \int_S \bar{u}^t [\partial^t n]\sigma\, dS - \int_V \bar{u}^t \partial^t \sigma\, dV \qquad [4.18]$$

This identity is in effect a three-dimensional form of integration by parts and can be applied to any two functions, and not just to σ and $\bar{u}$ as here. It is essential that the two derivatives are finite and continuous in the region V. We later apply the PVD to a volume divided into finite elements where, across the interface between elements, we recognise that the stress field σ may be discontinuous and, therefore, the derivative $\partial^t\sigma$ is not defined. However, we leave this possibility to later and now concentrate on what the PVD is delivering. So taking the framework argument as an example, we need to convert the term $\partial\bar{u}$ in the PVD to the form $\bar{u}$ and then appeal to the virtual nature of $\bar{u}$. Thus, substituting the Gauss identity into the PVD it becomes:

$$\int_V \bar{u}^t [\partial^t \sigma + p_V]dV - \int_S \bar{u}^t [(\partial^t n)\sigma - p_s]\, dS = 0 \qquad [4.19]$$

Now we use the same argument as for the framework. The displacements $\bar{u}(x, y, z)$ are virtual, so their coefficients must vanish, i.e.

$$\partial^t \sigma + p_v = 0$$

inside V and

$$(\partial^t n)\sigma = p_s \qquad [4.20]$$

on S.

These are the equations of equilibrium inside the body and on its surface. If we examine the first equation by expanding the term in $\partial^t\sigma$ and p, we find we get three equations representing the stress gradients, in the directions x, y and z, which equilibrate the three body force components if they are present. The second equation takes the stresses at a point on the surface, and then the components in directions x, y and z ($\partial'n$ are direction cosines) which have to equilibrate the components of the surface pressures p_{vx}, etc. We can therefore now rely on the PVD to satisfy equilibrium everywhere. The next step is the FE approximation.

4.4   Finite element approximations

Imagine the variation of displacement across a complex structure such as an aircraft or a cooling tower, as shown in Fig. 4.5. No simple theory can deliver this. The idea then is to divide the structure into elements which are small enough so that the displaced *shape* can be assumed with little error, and only the *magnitudes* of the displacement need be found. A crude approximation could be a linear assumption.

The whole structure can be divided into elements, the elements being either 1D (bars), 2D (plates) or 3D (bricks). Figure 4.6 shows a picture of the true displacement along a row of elements in 1D, and a linear approximation which will be quite close to the truth since the elements are quite small. Modern pre-processors use very powerful (yet friendly) algorithms to divide up the most complex shapes. Figure 4.7 shows a display of hexagonal brick elements where the elements are shrunk so that the user can check whether any are missing.

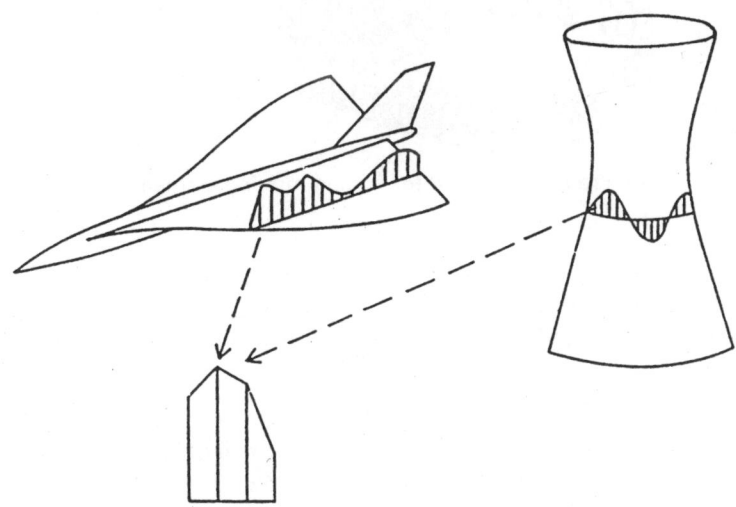

4.5 Two real structures.

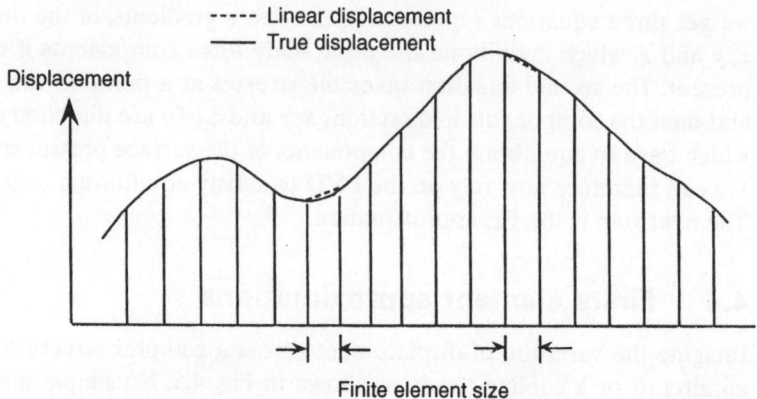

4.6 The 'finite' element approximation.

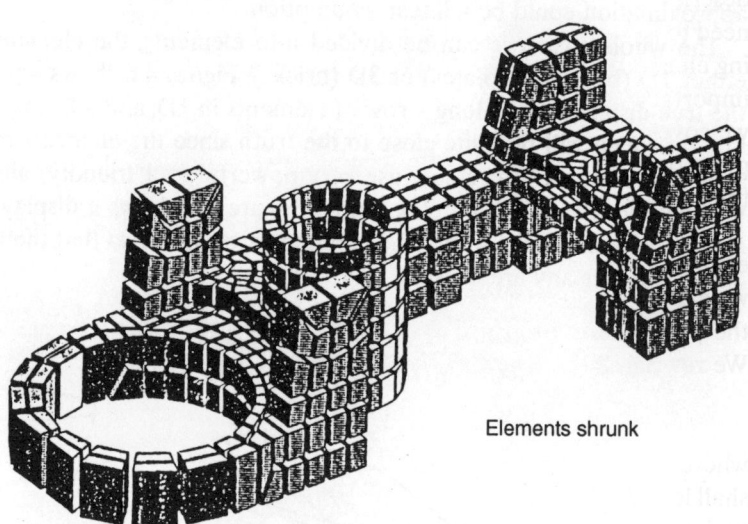

4.7 Pre-processor aid for finding missing elements.

Because we are using the PVD which is an integral, we can always write complete volume and surface integrals as a summation of all the element integrals:

$$\sum_{\text{elements}} \int_{V_g} \boldsymbol{\sigma}^t \bar{\boldsymbol{\varepsilon}} \, dV = \sum_{\text{elements}} \left[\int_{S_g} \mathbf{p}_s \bar{\mathbf{u}} \, dS + \int_{V_g} \mathbf{p}_v^t \bar{\mathbf{u}} \, dV \right] \qquad [4.21]$$

where S_g and V_g denote the surface or volume of the gth element.

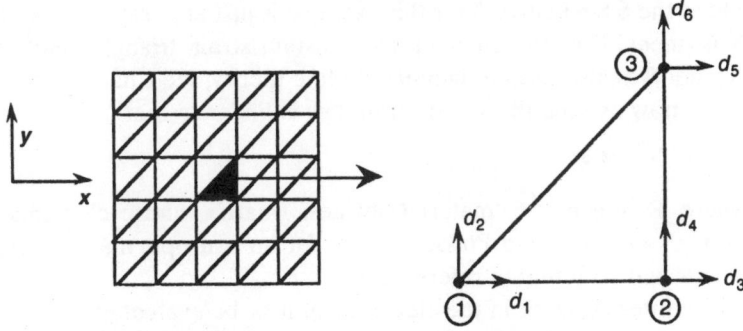

4.8 A triangular element with linear shape functions.

Now, it is not the intention in this book to delve deeply into the subtleties of the FE method, and individual elements in particular. It will suffice to describe the basic idea and mention particular aspects that the reader may need to know about when reading a manual, understanding the code, choosing elements and appreciating which errors are unimportant and which are important, and how to reduce or avoid them. For simplicity, then, suppose we cover a flat plate with a triangular mesh as in Fig. 4.8, and suppose the plate is in a 2D stress field (σ_{xx}, σ_{yy}, σ_{xy}) with uniform stresses across the thickness at any point.

The geometry is defined by the position of the three nodes of a typical element. If we *assume* that the displacement field $\mathbf{u}^t = [u, v]$ is *linear* then we can write $u = a + bx + cy$ and find the coefficients a, b and c in terms of the three horizontal displacements d_1, d_3 and d_5 at the three corner nodes. We rewrite the resulting expression as

$$u = N_1 d_1 + N_2 d_3 + N_3 d_5 \qquad [4.22]$$

where the 'shape functions' $N(x, y)$ are linear in this case. In Chapter 5 we shall look at more ambitious shape functions, such as quadratic, where more nodal points are required to define the field. The actual algebraic details of N_1, N_2 and N_3, which, for N_1 say, needs to be unity at node (1) and zero at nodes (2) and (3) can be found in reference 1. Similarly for the other displacement component:

$$v = N_1 d_2 + N_2 d_4 + N_3 d_6 \qquad [4.23]$$

For brevity we write

$$\mathbf{u} = \mathbf{N} \mathbf{d}_g \quad \text{where} \quad \mathbf{d}_g^t = [d_1 \ d_2 \ d_3 \ d_4 \ d_5 \ d_6] \qquad [4.24]$$

for the gth element.

It is now routine to go to the next stage and apply compatibility:

$$\varepsilon = \partial \mathbf{u} = \partial \mathbf{N} \mathbf{d}_g = \mathbf{B} \mathbf{d}_g \qquad [4.25]$$

where the 6×6 matrix **B** for this example is just an array of constants since **N** is linear. Thus the element is a constant strain triangle. (Note that the symbol **B** is also used in laminated plate theory; see Chapter 3.)

We now assume the stress–strain law is linear so that

$$\sigma = E\varepsilon \qquad\qquad [4.26]$$

where **E** is a 6×6 'material stiffness' matrix whose elements contain Young's modulus and Poisson's ratio for an isotropic material. Again the details will be found in reference 1.

The internal work in the element can now be evaluated

$$\int_{V_g} \sigma^t \bar{\varepsilon}\, dV = \int_{V_g} \varepsilon^t E \bar{\varepsilon}\, dV = \int_{V_g} d_g^t B^t E B \bar{d}_g\, dV \qquad\qquad [4.27]$$

But d_g can clearly be taken outside the integral, which itself can be evaluated since **B** and **E** are known.

Thus the element work = $d_g^t k_g \bar{d}_g$, where

$$k_g = \int_{V_g} B^t\, EB\, dV$$

is known as the *element stiffness matrix* since the work product may be viewed as $d_g^t k_g \bar{d}_g = (k_g d_g)^t \bar{d}_g =$ forces $\times$ displacements and force = stiffness $\times$ displacement.

The right-hand side of the PVD similarly becomes

$$\int_V p_v^t \bar{u}\, dV + \int_S p_v^t \bar{u}\, dS = \int_V p_v^t N d V \bar{d}_g + \int_S p_s^t N d S \bar{d}_g \qquad\qquad [4.28]$$

and these integrals can also be evaluated once we know the distributions of p_v and p_s. This external work can also clearly be written as $P_g^t \bar{d}_g$ where we see that

$$P_g = \int_{V_g} N^t p_v dV + \int_{S_g} N^t p_s dS \qquad\qquad [4.29]$$

must clearly be nodal forces. They are not real concentrated forces (which would produce infinite stress concentrations) and are often known as 'kinematically equivalent' forces since they are defined in terms of the displacement shape functions **N**. The above expressions for these nodal forces simply emerged naturally from our PVD and the element integrals. They are measures we find convenient to use. We see, therefore, that by assuming the displacement shape functions we have been able to evaluate the integrals for *any element*, and therefore for all elements. The FE method can be viewed as a way of numerically integrating. In fact the integrals are never done exactly, but as a summation of discrete values of the integrand times the appropriate bit of volume or surface at the sampling points. It is found that it is best to sample $\mathbf{B}$ (if it varies) at the optimal 'Gauss' points; see reference 1 for where these points are.

We have now converted the internal and external virtual work to simply $\mathbf{d}_g^t \mathbf{k}_g \overline{\mathbf{d}}_g$ and $\mathbf{P}_g^t \mathbf{d}_g$ and it only remains to sum

$$\sum_{\text{elements}} \mathbf{d}_g^t \mathbf{k}_g \overline{\mathbf{d}}_g = \sum_{\text{elements}} \mathbf{p}_g^t \mathbf{d}_g$$

In the simplest case, if all element displacements $\mathbf{d}_g$ are aligned with the global axes, they will be able to be selected from the total *global list* of displacements, usually written as

$$\mathbf{r}^t = [r_1 \; r_2 \; r_3 \ldots r_n] \tag{4.30}$$

where n may be many thousands. The total virtual work of the entire system, internal and external, may be similarly written as $\mathbf{r}^t \mathbf{K} \overline{\mathbf{r}}$ and $\mathbf{R}^t \overline{\mathbf{r}}$, where $\mathbf{K}$ is the whole structure's stiffness matrix and $\mathbf{R}$ is the global list of kinematically equivalent forces. But these two must be equivalent to the previous summations, so if we (or the code) make sure that we relate each $\mathbf{d}_g$ to the correct parts of $\mathbf{r}$, and the associated elements in $\mathbf{k}_g$, we may symbolically assemble

$$\mathbf{K} = \sum \mathbf{k}_g \quad \text{and} \quad \mathbf{R} = \sum \mathbf{P}_g$$

The exact details of how this is done methodically, and what transformations to $\mathbf{k}_g$ are necessary when element local axes are different from global, can be found in reference 1.

Having assembled both $\mathbf{K}$ and $\mathbf{R}$ the PVD can be rearranged as

$$(\mathbf{r}^t \mathbf{K} - \mathbf{R}^t)\overline{\mathbf{r}} = 0 \tag{4.31}$$

and we now use the argument that all the n components of $\overline{\mathbf{r}}$ are virtual so their coefficients vanish, and we have n equations

$$\mathbf{Kr} = \mathbf{R} \tag{4.32}$$

This set of equations is then solved as discussed in Section 4.7 whence, knowing $\mathbf{r}$, we then select the particular values $\mathbf{d}$ for the element level and hence $\varepsilon = \mathbf{Bd}_g$ and $\sigma = \mathbf{E}\varepsilon$ for the strains and stresses. This is the basis of the modern FE method. We see that the nature of the displacement field has been approximate, but that compatibility and the stress–strain law have been satisfied exactly. The use of the PVD to satisfy equilibrium therefore means that these equilibrium equations will be satisfied only approximately. It is therefore time to examine what this means.

We note that, by equating local and global nodal displacements, continuity of the structure's nodal displacements is assured. But for our triangular element the displacement variation between nodes is linear and

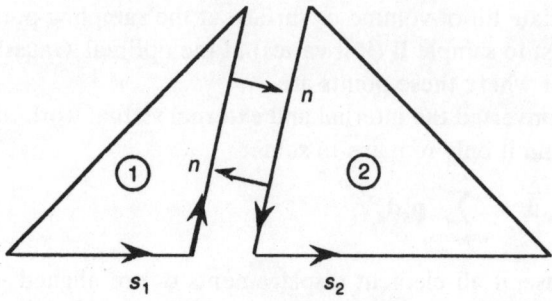

Two adjacent elements

4.9 Discontinuous stresses across an element interface.

therefore continuity across the entire element interfaces is also secured. Thus we have a continuous displacement field variation throughout the entire structure. However the strains, and therefore the stresses, are constant in any element, but the stresses will not be continuous from one element to another. Is this disastrous?

Consider two triangular elements (1) and (2) adjacent to each other as shown in Fig. 4.9. If we return to the general form of the PVD after transforming it, we can examine the consequence of summing all element contributions, and look at the coefficients of the virtual displacement across the interface between (1) and (2). If this face is not a loaded surface ($p_s = 0$) then we are grouping two contributions together with a common virtual displacement, i.e.

$$\int_S \bar{\mathbf{u}}^t[(\partial^t n)\sigma_1 - (\partial^t n)\sigma_2]\mathrm{d}s = 0 \qquad [4.33]$$

where the minus sign arises because the normal 'n' is in opposite senses for each element. Thus we may not be satisfying equilibrium exactly at any point, that is

$$(\partial^t n)\sigma_1 \neq (\partial^t n)\sigma_2$$

but it will be satisfied *in the mean* by making the above integral zero. Or, to put it another way, the out-of-balance error in stress components will do no net work over the displacements along the edge.

Moreover, if the element shape functions are capable of representing a rigid body motion (u = constant, v = constant, and the rotation), we recall that the equation of equilibrium is always associated with the corresponding virtual displacement, so the term like $\bar{u}$ = constant will ensure that the element equilibrium is satisfied in the x-direction. The rotation freedom will ensure that there is no resultant moment on an element if there are no body

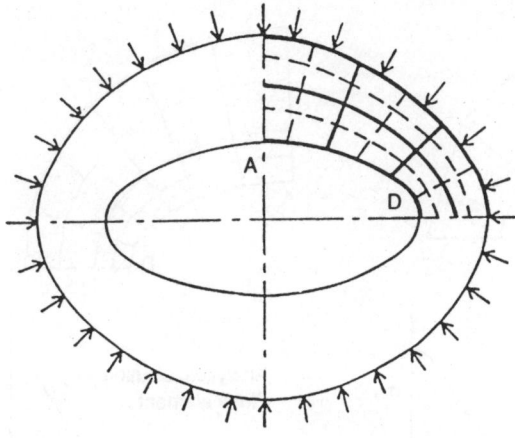

Elliptic 'annulus'

4.10 A stress concentration at point 'D'.

or surface forces acting upon it. We therefore accept that the FE solution only satisfies equilibrium *in the mean* and there will be discontinuities across interfaces. This does not mean that constant strain triangles, for example, should not be used. They are popular in describing, for example, elasto-plastic deformations since the whole element goes plastic and the integrals are easy. However, small elements may be needed to pick up rapidly varying strain fields.

Figure 4.10 shows a test case for an elliptical hole in a plate where there is a very high stress concentration at the point D. The plate is divided into four-node quadrilateral elements and in one quarter there is a coarse mesh of six elements and a fine mesh of 24 elements. In Fig. 4.11 it is clear that three elements around the edge are insufficient. If the mesh is refined then the discontinuities are much smaller and the maximum stress is predicted to within a few percent. The 'jumps' in the stresses can be used as an indicator of the error, and many codes then employ this measure to refine (or adapt) the mesh automatically until the errors are uniformly small everywhere. Most codes will take an average value of the stresses at a node coming from several elements sharing this node. If this is done in Fig. 4.11 we can see that the results will be quite good; except, of course, for the maximum stress at point D which sits astride the axis of symmetry.

We could further refine the mesh shown on the right of Fig. 4.11. It can be shown that if the element is capable of having a constant strain field, then successive mesh refinement will converge on the exact solution. Any element not capable of representing a constant strain field and a (strain-free) rigid body displacement is not acceptable. Another way to improve

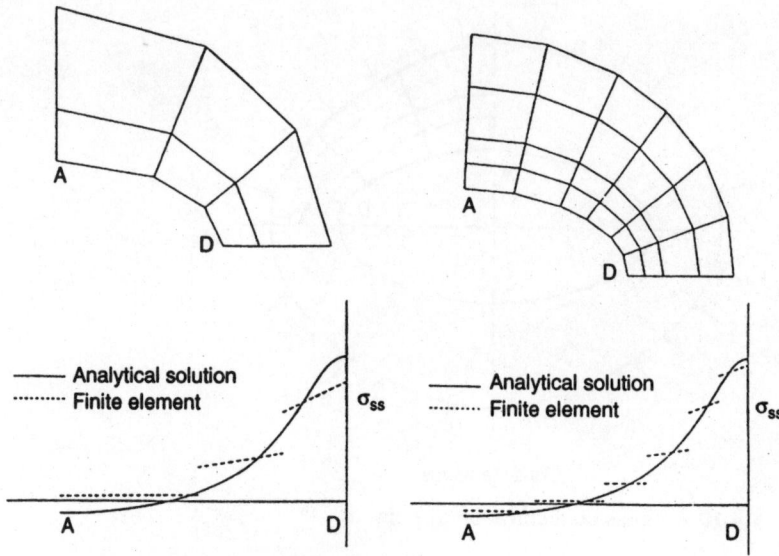

4.11 Effect of halving element sizes.

the description is to use higher order elements which use (say) quadratic displacement fields and twice as many nodes per element.

4.5   Brick, beam, plate and shell elements

The previous example of a triangle element, usually known as a membrane element, was selected for simplicity as an introduction to the finite element process. It is not too difficult to extend this concept to four-sided ('quad') elements with four nodes and eight nodal displacements, or eight- (and nine-) noded quads with mid-side nodes (and a centre node). It is also possible to design curvilinear elements better able to cope with curved boundaries. The trick here is to 'map' the shape of the element edges in terms of the nodal coordinates in exactly the same way as the displacement field was 'mapped' in terms of the nodal displacements. Such 'isoparametric elements' are a very powerful concept. They also have the property that, if the shape and displacement fields are mapped with respect to a reference unit square using non-dimensional shape functions, any rigid body movement still maps into zero strain fields.

The idea for 2D plates readily converts to 3D by deploying tetrahedral or hexagonal 'bricks'. A six-sided brick can have eight corner nodes and three displacement freedoms at each node for example. Higher order 20-node bricks are popular, having 8 corner nodes and 12 mid-side nodes. Both the displacement field and edges of such isoparametric elements can rep-

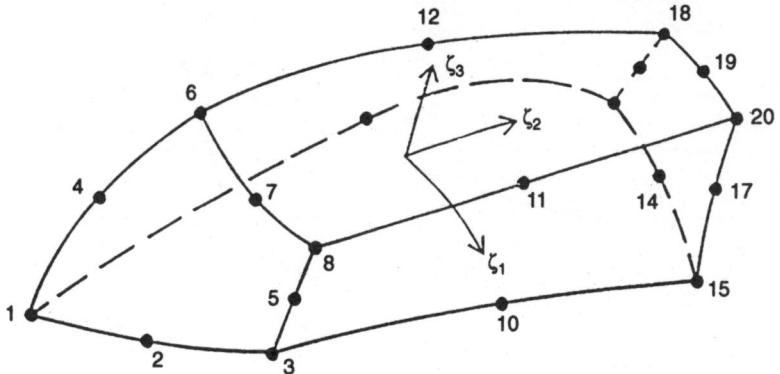

4.12 A 20-noded curvilinear brick element.

resent curved quadratic shapes. Figure 4.12 shows a curvilinear brick which will have $20 \times 3 = 60$ displacement degrees of freedom in total.

The variables ζ_1, ζ_2 and ζ_3 indicate that the shape in the x, y, z space has been mapped into the space $-1 < \zeta < 1$ which introduces the *Jacobian*, **J**, into the element stiffness.

$$\mathbf{k}_g = \int_{-1}^{1}\int_{-1}^{1}\int_{-1}^{1} \mathbf{B}^t\mathbf{EB}|\mathbf{J}|\,d\zeta_1 d\zeta_2 d\zeta_3 \qquad\qquad [4.34]$$

All the shape functions **N** are now functions of ζ_1, ζ_2 and ζ_3 and their contributions to the integral, via $\mathbf{B}(\zeta)$, is sampled at the Gauss points for all elements, the only varying quantity being the Jacobian matrix whose elements are the derivatives $\partial x/\partial \zeta_1$, etc.

A word should be said here about *distorted* elements since if bricks, and other isoparametric elements, differ too much from the basic cube then accuracy suffers. One way of sampling excessive distortion is the variation in the values of $|\mathbf{J}|$ at the Gauss points; many codes will flag warning messages. It is however possible to have undistorted elements with no $|\mathbf{J}|$ variations but having a very high aspect ratio, that is the element is long and thin in one direction. If the strain variation in this direction happens to be gentle than no problems will arise, but in general this will not be so. The ability of the shape functions to cope with variations in strain in all directions then becomes fully stretched, and many codes have empirical measures to flag warnings for excessively long elements.

Brick elements are expensive and produce a stiffness matrix with a high bandwidth. It is very easy to run out of computer memory when using them indiscriminately. For example, 100 elements in a row would not be unusual for a 1D or 2D problem, but if a 3D problem was modelled by $100 \times 100 \times 100$ 20-noded bricks the degrees of freedom escalate to 60 million!

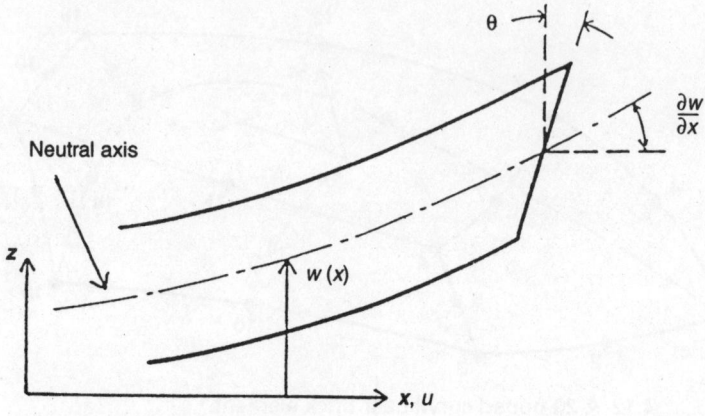

4.13 Slender beam bending.

However, many composite structures are thin-walled shells and plates which are nevertheless expected to react loads or pressures applied to their surfaces, i.e. they have to bend. Such structures are amenable to the assumption that the bending strains vary linearly through their thickness and any in-plane (membrane) loading leads to strains which are constant through the thickness. The shell or plate bending element, therefore, has a ready-made strain field assumption through the thickness so that we only need to describe variations over the plate or shell mid-surface. The simplest bending element is of course the 1D beam shown displaced in Fig. 4.13.

The standard 'beam theory' assumption is that 'plane sections remain plane' and in the figure the unstrained neutral axis is shown displaced by $w(x)$, and a vertical plane, originally normal to it, has rotated clockwise by $\theta(x)$ about the y-axis. Remembering that displacements are very small, we may approximate the distance above the neutral axis as z and write

$$u = z\theta$$

The application of compatibility then produces

$$\varepsilon_{xx} = \frac{\partial u}{\partial x} = z\frac{d\theta}{dx}$$

and

$$\varepsilon_{xz} = \frac{\partial u}{\partial z} + \frac{\partial w}{\partial x} = \theta + \frac{dw}{dx} \qquad [4.35]$$

In most cases where beams are slender we find that shear strains are negligible. We cannot get away with this in laminated composites because the shear modulus, mainly because of the matrix, is an order of magnitude

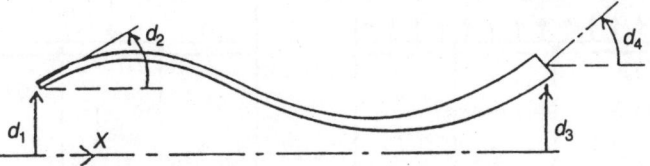

4.14 Cubic beam element freedoms (for continuity of displacement and slope).

smaller than the flexural modulus, which is mainly due to the much stiffer fibres. However, for an isotropic material we can neglect shear strains in a slender beam, i.e. $\varepsilon_{xz} \ll \varepsilon_{xx}$. If we then put $\varepsilon_{xz} = 0$, then $\theta = -dw/dx$ and

$$\varepsilon_{xx} = -z\frac{d^2w}{dx^2} \qquad\qquad [4.36]$$

This is the starting point for generating a beam element stiffness. We now have to choose nodal displacements, and clearly we need continuity of both w and θ, so the coventional degrees of freedom for a beam element are the displacements and rotations at the ends of the (say) two-node element shown in Fig. 4.14.

A cubic polynomial in x is completely defined by these four quantities and the cubic shape functions $\mathbf{N}$ are given in reference 1, as are the consequent beam stiffness and kinematically equivalent loads. If the displacement shape function is cubic, the bending strain given above is clearly linear. This beam element is very accurate because a cubic is a good high-order fit. In fact if the beam rigidity EI is constant, and the beam is loaded only at chosen nodes, the cubic displacement is exact.

The two solutions for a three-element clamped beam, loaded in two different ways, are shown in Fig. 4.15 where the plots could be bending strains or moments. Both solutions are approximate. The first solution has a distributed load and the second solution has a concentrated load (foolishly) applied in the middle of an element. In the first case the linear distribution clearly takes a mean path (in fact a least-squares fit), while in the second case the mean path for the middle element has to attempt to cope with a discontinuous variation. The FE method tries its best!

If the beam is made of a composite material with high stiffness fibres but low shear stiffness matrix, then we can no longer put the shear strain ε_{xz} to zero, so it is necessary to treat the displacement $w(x)$ and rotation θ separately. If a two-node element is still desired then we have to expand both θ and w as linear variations, and this leads to a constant bending strain along the element and a shear strain which varies linearly (but constant through the depth of course). This formulation is known as a Mindlin element. Prob-

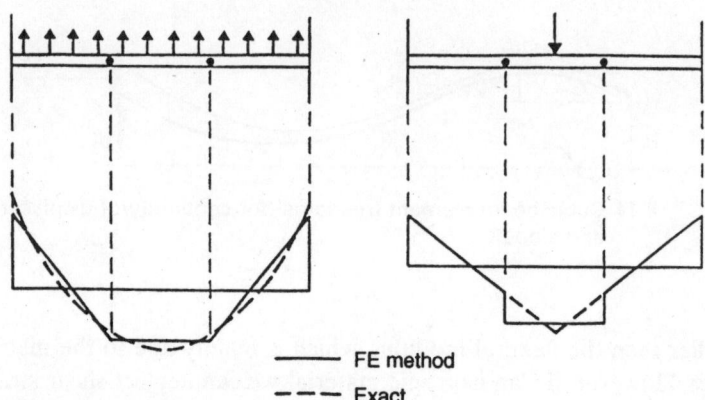

4.15 Clamped beam: two loadings.

ably a more acceptable version is the next higher order element with a mid-length node and quadratic variation in w and θ.

Flat plate elements are best thought of as 2D versions of beams, with the same assumptions regarding plane sections remaining plane. If we attempt to extend the conventional beam element it proves to be more difficult. It is easy to capture the bending moment variations, but a constant twisting moment is not straightforward, and the fundamental tenet of good finite elements is that all constant strain states should be represented to ensure convergence as the mesh size is refined. There is a host of plate bending elements with various mixtures of displacement and rotational freedoms, but probably the most popular has become the 2D version of the Mindlin beam element. It is straightforward to use and the only point to watch is its behaviour for very thin plates as the shear strains tend to zero. This forces the relationship $dw/dx = \theta$ and if these freedoms are no longer independent the determinant of the stiffness matrix tends to zero (see Section 4.7).

Thin shells have probably generated more academic papers in FE technology than any other topic. This text is not the place to raise all the issues and all the solutions. Most codes now have a degenerate or special form of the 60 degree of freedom brick element. Apart from the expense, it is not safe simply to use a thin brick as shown in Fig. 4.16(a) since the displacements through the thickness at any location will all be of the same order (particularly if through-thickness strains are negligible) and this leads again to the ill-conditioning mentioned in Section 4.7.

We therefore condense the system to that in Fig. 4.16(b) with only eight nodes along the mid-surface $\zeta_3 = 0$, at which all three freedoms (u, v, w) are used. The remaining freedoms will be the two rotations of the normal which have to represent the outer surface freedoms, and which have been removed from Fig. 4.16(a). This can be done by adding a linear variation

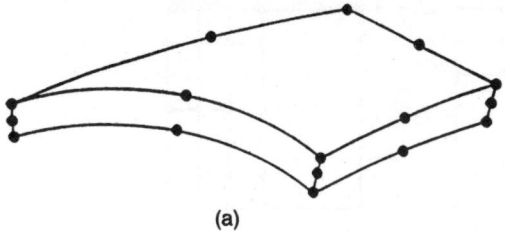

(a)

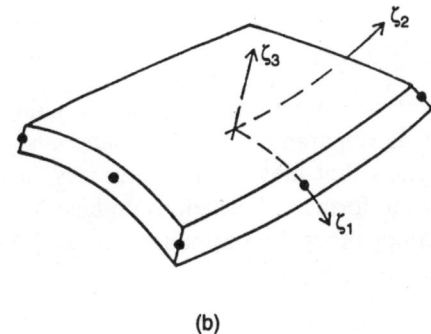

(b)

4.16 (a) A brick element 'thinned'. (b) A 'respectable' thin shell element.

(with ζ_3 of course), but other methods deploy a simplifying integration scheme on the 'brick functions' which amounts to the same thing.

4.6   Model generation

All commercial codes have a pre-processor, or will hook onto another commercial pre-processor, to generate the shape and the meshes. These pre-processors are becoming very sophisticated and user-friendly, and indeed the need for human intervention between a computer-aided design (CAD) description and the FE model is fast disappearing. The need for the user to tailor the mesh to cope with areas of stress concentration is also theoretically disappearing with the emergence of adaptivity routines which will reduce element size (h refinement) or increase the order of the elements (p refinement), or both. Most users still prefer to control the generation of the model, probably by interactive graphics-driven manipulation, and choosing elements' material properties, etc., by picking off from a supplied database. Very briefly the model will be generated in the following order:

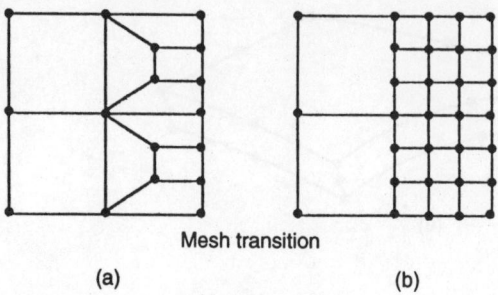

Mesh transition

(a)   (b)

4.17 Two ways of grading a mesh.

4.6.1 Geometry specification

If a distinct 'region' is generated, and bounded by splines say, then the code can subdivide the region using isoparametric contour intervals, so the user needs only to generate regions that may be lines, surfaces, or solids, the latter often being swept by the former. The regions do have to join seamlessly so that artificial cracks or other discontinuities are avoided. Various checks and blending routines are necessary.

4.6.2 Mesh generation

Having generated regions, and chosen the mesh density in each region, the user may have to choose or dictate some refinement. Traditional ways of ensuring all adjacent elements have common nodes are obvious and shown in Fig. 4.17(a). However, the transition elements shown are quite a departure from the desirable square, and may be even more distorted if the real curvilinear region is mapped onto Fig. 4.17(a). A much easier way is to drop nodal continuity as shown in Fig. 4.17(b), and allow the code to ensure continuity of displacements by forcing the refined element nodal displacements to follow the displacement pattern along the edges of the coarse element. This 'multipoint constraint' enforcement is available in most codes.

4.6.3 Element types

We have already discussed the types and the option of low- and high-order elements, but the user still has to decide whether the internal stresses can be represented by a 1D beam, or 2D plate, or pure membrane shell, a 3D shell with bending, or a general 3D solid. It is at this stage that the cost of the model is decided and it is still true that 'engineering judgement' is needed. It is an expensive mistake to use 3D bricks when in doubt. The modelling is particularly important (and difficult) in layered composites when we would like to use 2D elements but the threat of through-thickness stress variations and local stress concentrations is ever present. Various strategies for mimicking 3D bricks, to get through-thickness shears and peeling stresses (but reducing the element stiffness to compete with 2D plate and shell elements), are discussed in Part III.

4.6.4 Material properties

For isotropic or orthotropic materials the user will simply be asked for moduli and Poisson's ratios which occur in the aforementioned **E** matrix. For composites the choice is much wider and this aspect is dealt with later.

4.7   Solution procedures

In general solution procedures should not concern the user unless he/she is asked to choose, or unless the procedure fails! Earlier we stated that the equations $\mathbf{Kr} = \mathbf{R}$ have to be assembled and solved, so a few points need to be made. Firstly any structure needs to be adequately supported so boundary conditions need to be specified. Some displacements (including rotations) will need to be set to zero. This could also include axes of symmetry if only a half (or a quarter, etc.) of the structure needs to be modelled. The removal of such displacements from **r** is always done after the unsupported stiffness matrix has been assembled, and consequently the corresponding rows and columns are deleted. Before this happens the equations could not in fact be solved because an unsupported structure cannot resist loads. Displacements could occur without any applied loads or internal strains, i.e. there would be a solution to $\mathbf{Kr} = \mathbf{0}$.

If such a solution is possible, **K** is said to be singular and there are several purely mathematical ways of testing this. The determinant, for example, is zero. All element stiffness matrices are singular for the same reason, but only if they are then not assembled with the correct connectivity will there be a problem. In practice the determinant check is not realistic since a perfect 'zero' will never be the outcome. Moreover some structures will just be poorly supported, for example, and then the determinant will just be 'small'. The determinant of the set of equations in **r** or **d** will be zero if there exists a linear dependency between the displacements (like the rigid body movement), but this can happen as mentioned previously in Mindlin elements for example when the shear strains become small. Integration tricks can be deployed selectively by using reduced integration on the stiffness terms arising from the shears. Other forms of 'locking' can arise and many codes will flag when and where this occurs. A quantitative measure, known as a conditioning number, is often quoted. The ratio of maximum to minimum eigenvalues is one such measure, but is expensive to evaluate for very large systems. A better way is for the errors to emerge during the solution procedure and this also brings out the effect of the computer's word-length and rounding errors.

When we speak of solving $\mathbf{Kr} = \mathbf{R}$ it should be said that $\mathbf{K}$ is never inverted. The number of degrees of freedom can run into hundreds of thousands and the formal inversion would be extraordinarily expensive. Two solution procedures are used: direct and indirect (iterative) methods. The direct method is best able to exploit the fact that the stiffness matrix $\mathbf{K}$ has a *finite* bandwidth, with most off-diagonal terms being zero. This may not happen naturally during assembly and will depend on the way the regions and elements are created. There are standard element or node renumbering schemes to ensure that the elements of $\mathbf{K}$ are grouped as closely as possibly to the leading diagonal.

Two direct methods are used, which decompose or factorise $\mathbf{K}$ into products of two triangular matrices. Gaussian elimination is used in conjunction with the frontal solution method (after element optimum renumbering) when the assembly and factorisation proceed in parallel so that the full set of equations is never assembled. The other closely related technique is the Cholesky factorisation which exploits the fact that $\mathbf{K}$ is symmetric and positive definite. The complete matrix is factorised and this time it is assumed that node renumbering has been optimised. During the course of the factorisation, if the equations are ill-conditioned, the size of the diagonal terms becomes unacceptably small. One widely used conditioning number is the 'trace' (i.e. the sum of the diagonal terms) divided by the lowest term. The order of magnitude of this number should be less than half of the number of significant digits in the mantissa of the computer word-length. Another measure is 'diagonal decay', i.e. the ratio of a leading diagonal term after triangularisation to that of its value before.

Most workstations use 64-bit processors and natural ill-conditioning is not really a problem nowadays, but commercial codes will have some version of a conditioning number, or will stop the solution when the leading diagonal term becomes too small and inform the user. Sometimes a support or part of the structure will be excessively flexible and, if this emerges during the solution procedure, then most codes will flag the location to be checked.

It should be noted that the solution time of direct methods varies roughly as the cube of the problem size, and for large 3D problems, with very large bandwidth brick elements, it is now common to use *indirect* solvers where execution time can vary more like n^2. Indirect iterative methods, such as the conjugate gradient technique, exploit the fact that if we seek to satisfy equations of equilibrium any residual errors can be fed back and an improved solution sought. They rely heavily on the success of choosing a good initial vector $\mathbf{r}$ and much effort has gone into several pre-conditioning techniques. The success of such methods is judged by the rate of convergence and this depends on the nature of the structure, for example are there stiff regions and flexible regions where **Kr – R** residuals vary significantly? This is where the pre-conditioning comes in.

4.8   Results verification

Modern FE codes sell if they are user-friendly, and if the answers look convincing. There is a temptation therefore to conceal rather than eliminate errors. For example we noted that the Gauss points selected for numerical integration, in the element stiffness integrals, were optimum choices. These points are also optimal in sampling the strains, and hence stresses after solution, assuming that we are using a higher-order element rather than the constant strain variety. Most codes therefore fit curves through the Gauss points and extrapolate to the element nodes. This results in several values when more than one element shares the same node, and usually the arithmetic mean is taken. This can be very accurate, and is usually taken as the 'exact' value when estimating errors and adapting the mesh. However, such averaging can conceal large jumps and users should be aware that this has happened. Another point to remember is that most errors occur at the modelling stage, which is where experience counts. So how best to present the results and then be able to judge? Any subjective assessment is best backed by a second opinion and many company QA procedures insist on this.

The most obvious check is to try another, preferably finer, mesh and idealisation if you can afford it. Noting that the FE method based on the PVD will tend to underestimate displacements and strains (the solution is 'over-stiff') we expect the stresses to converge from below. The answers therefore are not conservative. We have mentioned reduced integration, whereby the number of Gauss points chosen is less than the optimum, which can be dangerous, but mostly 'softens' the element and is therefore beneficial as well as cheaper. Displacement plots, which show the exaggerated deflected shape presented against the background of the undeformed shape, are impressive and will certainly reveal gross, unlikely looking behaviour. They will not, however, reveal the same details as stresses and strains which depend on gradients of the displacement, in fact second derivatives in the case of some beam and plate elements as we recall. Automatic control of numerical accuracy, by adapting the first mesh according to the error distributions in the first solution, is nowadays driving commercial codes in the direction of a 'black box' with minimum user interference other than stating an acceptable tolerance. Adaptivity can be expensive, but the price of hardware and memory continues to plummet so automatic mesh refinement will be attractive for a structure where all the elements are of the same family. The user will be able to see the areas where refinement has taken place and be able to identify the source of the local stress concentration. Figure 4.18,

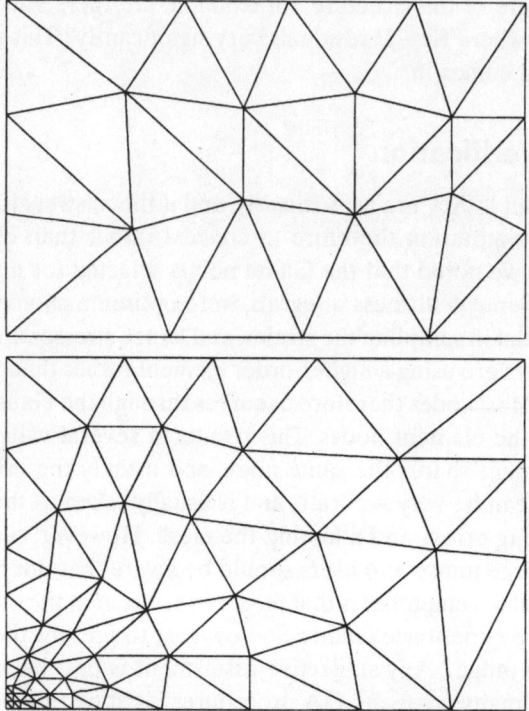

4.18 Adaptivity at work: refining a mesh near a stress singularity.

for example, shows the adaptive meshing happening for a plate with a concentrated force applied to the bottom left-hand corner, and thereby producing a singular behaviour.

'p' adaptivity, in which the extra degrees of freedom are achieved by raising the order of the displacement shape polynomials, will leave the original mesh alone, and the code needs to identify where the elements have been so treated. We are now at the stage where the mesh does not have to be displayed at all. The values of stress, strain, or equivalent measure, are evaluated at all sample points and then contours are displayed.

Selection of the stress to be displayed can be complex. A single equivalent stress such as that used in ductile failure criteria is an example (von Mises and Tresca), or if the material is brittle then the maximum principal stress or strain is often chosen. Probably a better choice would be stress intensity factor or energy release rate for brittle materials. Single scalar quantities are clearly easier to display, preferably as contours. However, in general there may be six stress components, and they can be in either local or global directions. In laminated composites we could be looking also at stresses in only one of many layers. Commercial codes have only just become willing to face this problem. Contours confined to a single lamina provide one approach, remembering that the lamina may be curved in a composite shell, for example. Laminated composites tend to fail through deficiencies in their through-thickness strengths and here it is feasible to drop a 'bore hole' through the laminate and present the variations along it. Remember that displacements are always continuous but stresses may not be, even when they should be. If laminae are parallel to the x–y plane, for example, then ε_{xx}, ε_{yy} and σ_{zz}, σ_{zx} and σ_{zy} should be continuous, but if the laminae have different stiffness properties then σ_{xx}, σ_{yy} and ε_{zz} will not be continuous. These discontinuities between the individual laminae are one reason why higher-order bricks cannot be used to capture the through-thickness stresses. Their continuous displacement shapes will produce continuous strains but discontinuous stresses whatever the component. Recent developments use 'zig-zag' shape functions which try to overcome this deficiency.

Finally, laminated composites are prone to delaminate, driven by the 'peel' tensile stress (mode I) or shear stress (mode II). The onset of delamination and its propagation is best predicted by the energy release rate (rather than stress) but evaluating this is still a current research topic and only a few commercial codes can presently do this. Failure criteria and many other topics peculiar to laminated composites are addressed in the following chapters.

4.9   Summary

The basis of the finite element method is absolutely straightforward. In each element a displacement field 'shape' is *assumed*, i.e. $\mathbf{u} = \mathbf{Nd}$.

Compatibility is then applied exactly

$$\varepsilon = \partial^t \mathbf{u} = \mathbf{Bd}$$

The stress–strain law is exact also

$$\sigma = \mathbf{E}\varepsilon$$

The PVD then satisfies equilibrium *approximately*, leading to an *element* stiffness matrix $\mathbf{k}_g$ and converting the applied loads to kinematically equivalent forces $\mathbf{P}_g$. The PVD is now summed over the entire structure which produces a global stiffness matrix $\mathbf{K}$ relating the global displacements $\mathbf{r}$ to the global forces $\mathbf{R}$.

The equations $\mathbf{Kr} = \mathbf{R}$ are solved, and on back-substituting $\mathbf{r} \rightarrow \mathbf{d} \rightarrow \varepsilon$ we recover the strains and stresses in every element.

Special shape functions are used for beams, plates, shells, solids, etc. Generating the model and solving for the displacements and stresses is made straightforward by most commercial codes. Poor meshes and FE errors can be assessed by looking at stress 'jumps' across element interfaces, or else by stipulating a tolerance and letting the code adapt.

Special requirements for composite elements, whether they be part of beams, plates or shells, are addressed in the next chapter.

4.10   Reference

1 *A Finite Element Primer*, National Agency for Finite Element Methods and Standards, East Kilbride, Scotland, 1986.

Part III
Finite elements applied to composite materials

D. HITCHINGS

This part of the book builds on the basics of FE analysis presented in Part II. In the first chapter consideration is given to the particular issues that arise when applying finite elements to composites, especially to the layered nature of the material. The second chapter addresses how composites should be defined in FE analysis.

Composites and finite element analysis

5.1   Elements used for composite laminate analysis

The layered nature of composite materials means that only certain element types can be used efficiently within the FE analysis of composites. It would, in theory, be possible to stack three-dimensional brick elements with one layer of bricks representing a ply of composite material. However, this is generally impractical for two reasons. It would be very expensive to run such a model if the lay-up had more than a few plies and a real structure was being represented. In addition, layering brick elements through the thickness of relatively thin plates leads to very ill-conditioned sets of equations. For these reasons bricks tend only to be used either where the composite lay-up is very thick and the geometry is more solid than plate like (not very common), or where there is a 3D stress field in the material, as can occur at the free edges of plates. In the latter case the 3D nature of the stress field is only significant over a short length, typically 3–5 thicknesses, and a small 3D sub-model of the region can be used. Note that these restrictions do not apply to the composite solid plate element discussed later.

Figure 5.1 shows a typical four-node quadrilateral (quad) flat plate element with only in-plane displacements. This is known variously as a 2D brick or a membrane (since it can only sustain stresses in the plane of the material) element. Such elements are widely used in composite analysis for modelling the in-plane behaviour of flat plates. However, care must be taken to ensure that both the stresses and the displacements are zero out of the element's plane, otherwise the model is not valid. With layered composites this depends upon the orientation of the plies. If the lay-up is not balanced then this 2D element cannot be used.

In practice it is much more usual to employ some form of plate or shell element. A typical nine-node shell element is shown in Fig. 5.2. The shell is modelled in terms of its mid-surface plane rather than the complete volume. The standard bending theory assumption that the strains only have a combination of constant and linear variation through the thickness of the shell

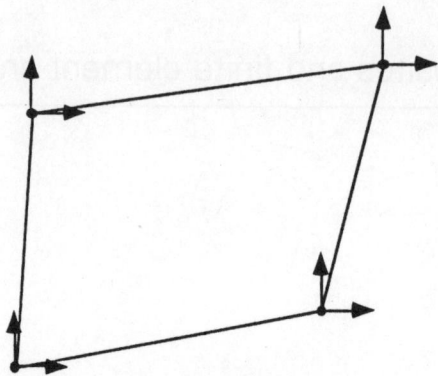

5.1 Four-node 2D solid element.

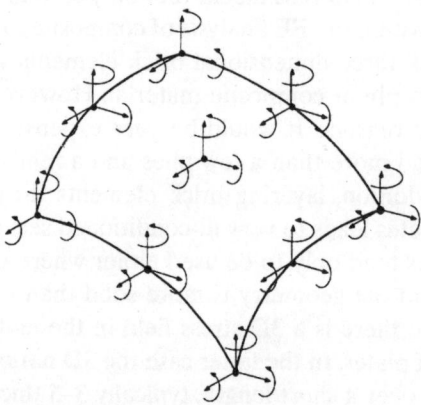

5.2 Nine-node Mindlin shell element.

is then used. This means that the deformation of the shell can then be defined by stretching (for the constant strain components) and rotation (for the linear strain components) of the shell's mid-thickness surface.

Both the two-dimensional solid and the general plate element can occur as quadrilaterals or triangles. They all have nodes at the corners of the element and some also have mid-side nodes. A few have internal nodes, as shown in Fig. 5.2, and this can be especially useful for improving the accuracy of the shell's bending behaviour. Generally, the more nodes on the element, the coarser the finite element mesh can be. For a mesh with a given number of nodes, the more nodes per single element then the more accurate the results will be.

Provided that the shell is thin then shell theory gives a very good description of its behaviour. It is assumed that there is no stress or strain in the direction normal to the plane of the mid-surface. With this idealised behaviour it is only necessary to define displacements and rotations as the degrees of freedom on the element mid-surface plane. However, if classical Kirchhoff thin shell theory is used it is found to be very difficult to derive finite elements for other than very simple (rectangular) geometries. This arises from the need to differentiate the transverse displacements twice to derive the bending strains. Definition of the shape functions and determination of the Jacobian matrix for arbitrary shaped elements has not proven to be simple in such cases. Instead, other shell theories have been used, typically the Mindlin theory. These allow for transverse shear strains to occur so that the bending strains take the form

$$\varepsilon_{x'x'} = \frac{\partial u'}{\partial x'} + z' \frac{\partial \theta_y}{\partial x'} \qquad [5.1]$$

where the 'prime' denotes that these are coordinates and displacements in local coordinates in the plane and normal to the mid-thickness surface. There are other similar terms for the other bending strains. This definition of strain only requires first derivatives to be found, which greatly simplifies the determination of the Jacobian matrix. In addition, the formulation allows for the transverse displacements w' and the rotations $\theta_{x'}$ and $\theta_{y'}$ to be interpolated independently of each other. It is this which greatly simplifies the definition of the shape functions. Standard shape functions of the form

$$w' = [N_1 N_2 N_3 \ldots N_m] \begin{bmatrix} w_1 \\ w_2 \\ . \\ w_m \end{bmatrix}$$

and $\qquad\qquad\qquad\qquad\qquad\qquad\qquad\qquad\qquad\qquad\qquad\qquad [5.2]$

$$\theta_{x'} = [N_1 N_2 N_3 \ldots N_m] \begin{bmatrix} \theta_{x1} \\ \theta_{x2} \\ . \\ \theta_{xm} \end{bmatrix}$$

can be used, where the shape function for the ith node, N_i, is the standard simple form used for many 2D membrane plate and brick elements.

The Mindlin theory allows shear strains to occur and so these must be also included in the element behaviour and take the typical form (constant through the thickness):

$$\varepsilon_{x'z'} = \frac{\partial w'}{\partial x'} + \theta_y \qquad [5.3]$$

In reality the transverse strains cannot be constant through the thickness since they must be zero on the top and bottom faces of the plate, and they will generally vary in some parabolic form through the thickness. The transverse shears in the Mindlin theory therefore represent average through-thickness values. For thin plates the consequence of the variation through the thickness is second order and this average value is quite sufficient for good accuracy. However, as discussed later, some care must be taken when using composite materials if a shear correction factor is employed to account for the through-thickness variation.

A typical Mindlin shell element is shown in Fig. 5.2. This is a quadrilateral element with nine nodes. There are six degrees of freedom at each node, three translations and three rotations. They act in the global coordinate directions, allowing the shell elements to be assembled in the usual manner. It is possible to define an eight-node version of the element, without the mid-face node, but for general shell bending the centre node is very beneficial in allowing the shape functions to better define the curvature of the shell. It has the same effect in giving a better description of the geometry; the eight-node quad formulation tends to flatten the geometry at the centre of the element so that, typically, a spherical shape is rather distorted.

For an element to be useful it must pass the 'patch test'. For a thin shell this means that it must reproduce a constant pure bending for any amount of element distortion. This is illustrated in Fig. 5.3 for the nine-node quad. The mesh of distorted elements (upper left window) models a simple square plate which is supported so that rigid body motions are just suppressed (upper right window). It is loaded by equal and opposite moments along the two edges parallel to the y-axis (middle left). The bending distortion is relatively complicated since it includes anti-clastic curvature (middle right). However, the direct x-stress is constant across the top face and the other stress components are zero (lower windows). The element passes the patch test.

One significant problem associated with the Mindlin element is that the basic formulation only requires rotational freedoms that bend the plate. If the structure were a single smooth surface then only the three translation and the two bending rotational freedoms (five in all) would be required. However, this smooth geometry is not generally the case and all three rotations have to be included, giving an element with six degrees of freedom per node. The third rotation is the in-plane rotation about the normal to the plane. It is often called the 'drilling' rotation and is shown in Fig. 5.4. This is an in-plane degree of freedom but it is not a natural freedom and is difficult to include properly within the in-plane strain definitions. In most FE systems it is treated as a fictitious freedom that does not couple to any of the other degrees of freedom on the element. It has to have some

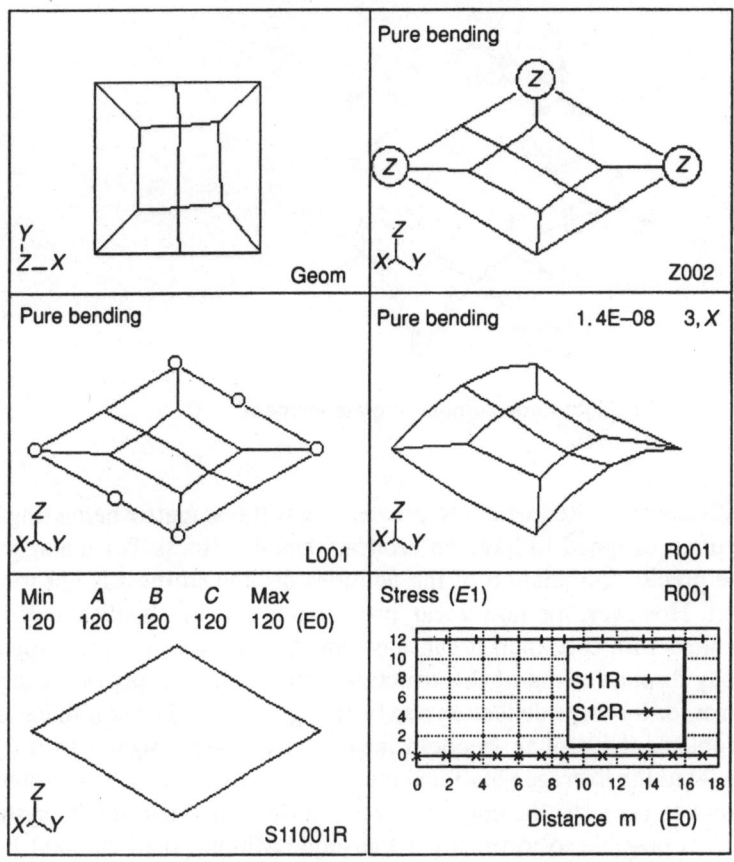

5.3 Patch test for nine-node Mindlin shell element.

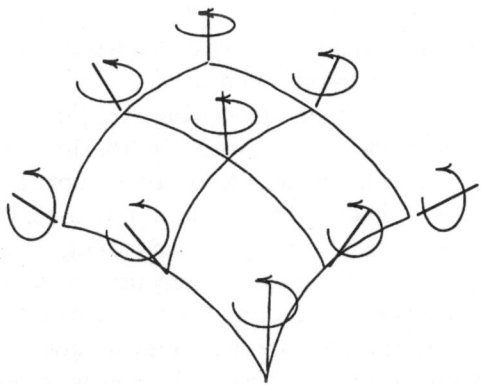

5.4 Drilling freedoms for the nine-node Mindlin plate element.

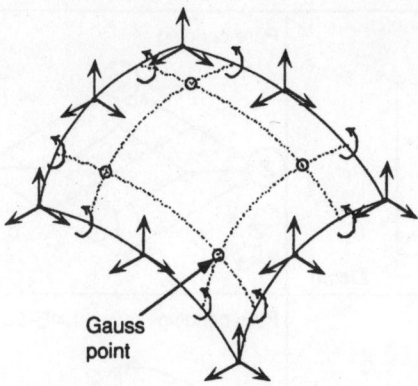

Gauss
point

5.5 Eight-node semi-Loof plate element.

stiffness associated with it to prevent the stiffness matrix being singular and
is often assigned to have an arbitrary small stiffness. For linear problems
the precise formulation of the fictitious drilling stiffness is not too important. However, for non-linear problems it can be significant. If it is not
defined with care then it can have an in-plane stiffness for in-plane rigid
body displacements of the structure, which can then provide a significant
error for the large deflection and buckling response. For non-linear dynamic
problems the drilling freedom must have a mass assigned to it (typically
equal to the average rotational inertia of the two out-of-plane rotations) in
order to preserve the correct inertias in the global coordinate system.

It is possible to formulate a four-node Mindlin shell element but some
care must then be taken with the numerical integration scheme used. The
transverse shear terms must be integrated to a lower order than exact integration would suggest to prevent the element from being significantly over-
stiff. This has the consequence that the element has some zero strain energy
non-rigid body displaced shapes, so-called 'hour glassing modes' and, if the
boundaries are not sufficiently supported, these modes can propagate
throughout the structure and lead to ridiculous results.

The problem of what to do about the drilling freedoms has been
addressed with the semi-Loof element, as shown in Fig. 5.5. The typical form
of this element has eight nodes with the three global translations as degrees
of freedom at the nodes. In addition it has two local rotational freedoms
along each edge and these are located at the points on the edge which inter-
sect with the lines through the Gauss integration points; the 'Loof' points.
This gives the element 32 degrees of freedom. Since the rotations are not
at corners, and only the local rotations are used, they do not need to be
transformed to global directions for assembly and the need for drilling freedoms does not arise. One obvious problem with this element is that some

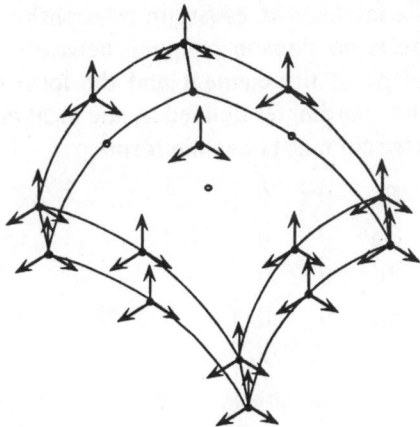

5.6 Eighteen-node 'solid' plate element.

of the degrees of freedom occur at points other than nodes and a non-standard assembly process is required. The semi-Loof element also includes the effects of transverse shear strains. Comparisons between the behaviour of Mindlin and semi-Loof elements indicate that they are comparable in accuracy and efficiency.

Another form of shell element is the 'solid' shell, as shown in Fig. 5.6. This has the geometrical definition of a solid element but the behaviour of a shell. The element has any number of nodes, depending upon the actual formulation, but only two nodes through its thickness. The element in Fig. 5.6 has nine nodes on the top surface, nine on the bottom and no mid-thickness nodes, giving an 18-node element. Each node has only translational freedoms so that no rotational freedoms are explicitly defined and there is no problem with drilling rotations. It has exactly the same number of freedoms as the nine-node Mindlin plate element of Fig. 5.2. The components of the translational displacements in the plane of the element define both the stretching and the bending of the element. The average value from a pair of nodes on the top and bottom surface is equivalent to in-plane stretching, and their difference divided by the shell thickness gives the mid-plane rotations.

The Mindlin shell and the 'solid' shell do not have exactly equivalent degrees of freedom. The 'solid' shell has no drilling freedom but it does have two through-thickness translational freedoms, allowing through-thickness direct strains (and stresses) to occur, which the Mindlin element does not have. It would be quite possible to use a standard brick element to model the shape but, if this were done, it is found that the conventional brick is over-stiff and the thinner the plate, the more over-stiff it becomes. To overcome this in the 'solid' shell a modified stress/strain relationship is used. Here it is assumed that there is no Poisson coupling between the local stresses in the mid-surface plane of the element and the local through-thickness stresses. Hence, in the coordinates defined by the local mid-plane directions, the material stress–strain matrix has the form:

$$\mathbf{E}' = \begin{bmatrix} E_{11} & E_{12} & 0 & E_{14} & 0 & 0 \\ E_{21} & E_{22} & 0 & E_{24} & 0 & 0 \\ 0 & 0 & E_{33} & 0 & 0 & 0 \\ E_{41} & E_{42} & 0 & E_{44} & 0 & 0 \\ 0 & 0 & 0 & 0 & E_{55} & E_{56} \\ 0 & 0 & 0 & 0 & E_{65} & E_{66} \end{bmatrix} \qquad [5.4]$$

The terms E_{11}, $E_{12} = E_{21}$, E_{22} are the in-plane direct stress/strain properties and have values corresponding to two-dimensional plane stress properties, E_{44} the in-plane shear modulus and $E_{41} = E_{14}$, $E_{42} = E_{24}$ are the in-plane shear coupling terms and are often zero. E_{33} is the through-thickness Young's modulus and this has no coupling to any other term in the material stiffness matrix. The remaining terms, E_{55}, $E_{56} = E_{65}$, E_{66} are the transverse shear terms and they do not couple to any other terms in the material stiffness matrix.

The 'solid' thick shell element then has a behaviour almost identical to that of the nine-node Mindlin element. The 'solid' shell has various advantages over the Mindlin element. It has no fictitious drilling freedom and elements can be stacked through the plate thickness if required. It does have one major disadvantage in that the through-thickness stiffness makes it rather more prone to rounding error in the solution process. It is also more cumbersome to model corners and other connections with the 'solid' shell. However, it does have the advantage of coupling directly to standard brick elements.

5.2   Accuracy of the finite element solution

The finite element is, by its very nature, only an approximate solution, as already explained in Section 4.4. It only guarantees that equilibrium is satisfied on average over an element. It does not satisfy equilibrium over any smaller volume than a complete element. Obviously, as the mesh size is decreased then equilibrium is satisfied over smaller and smaller patches. The art of using the FE method lies in designing a mesh that will deliver results to an acceptable accuracy but with a mesh density that is not prohibitively expensive to run.

The fineness of the mesh can be important for linear static problems but it is especially important for the analysis of non-linear and dynamic problems, where repeated solutions of the equations are required.

The significance of the mesh, and the corresponding loss of accuracy if the mesh is too coarse, can be illustrated by the analysis of a simple cantilever under its own weight (Fig. 5.7). The model is fully fixed at the cantilevered end. It has been modelled using four meshes, a 3×10 mesh of four-node quad elements, a 9×30 mesh of four-node quad elements and the same with eight-node quads. The results of these four analyses are shown in Fig. 5.7. Figure 5.7(a) shows the deflected shape and the direct stress distribution along the top edge of the beam for the eight-node quad elements. The results for these two meshes are almost identical. The peak displacement is 23 mm and the peak stress 48 MN/m^2. The agreement means that these two meshes have converged and the results are accurate. Figure 5.7(b) gives the results for the four-node quads. The coarse 3×10 mesh gives very poor results. The peak displacement is 8.9 mm (about one quarter of the converged results of the high order elements) and the peak stress is 17 MN/m^2. The finer 9×30 mesh of four-node quads is much closer to the converged values with 19 mm and 40 MN/m^2, but these are still significantly below the converged values.

The displacement assumption used in forming the FE stiffness matrix means that the finite element will be over-stiff, as mentioned in Section 4.8. The above results show that a coarse mesh of low-order elements is very over-stiff. The coarse mesh of higher-order elements gives a more flexible solution and the finer the mesh, the better the solution. This is generally the case and it is now standard practice to use a relatively coarse mesh of high-order elements for preference, rather than a fine mesh of low-order ones.

One point to note from this exercise is that very little can be read into the form of either the displacement or stress curves. Even when the error is very significant the general shape and pattern of the results are correct. When using the FE method there is always a danger that the mesh might not be fine enough. This can be best investigated by carrying out a series of runs with different mesh densities or different element types and checking for convergence. Some indication of the error in the solution can also be obtained by looking at stress differences across element boundaries, and the magnitude of stresses on external boundaries that should be stress free. However, the correct interpretation of these values does require some experience. Such interpretations have been automated in some systems in the form of error estimates (see Sections 4.7 and 4.8).

5.3   Other modelling techniques useful for composite analysis

Before discussing the specific application of finite elements to composite analysis some other aspects of FE modelling that are useful for composites will be illustrated. Composites are usually used as laminated materials and

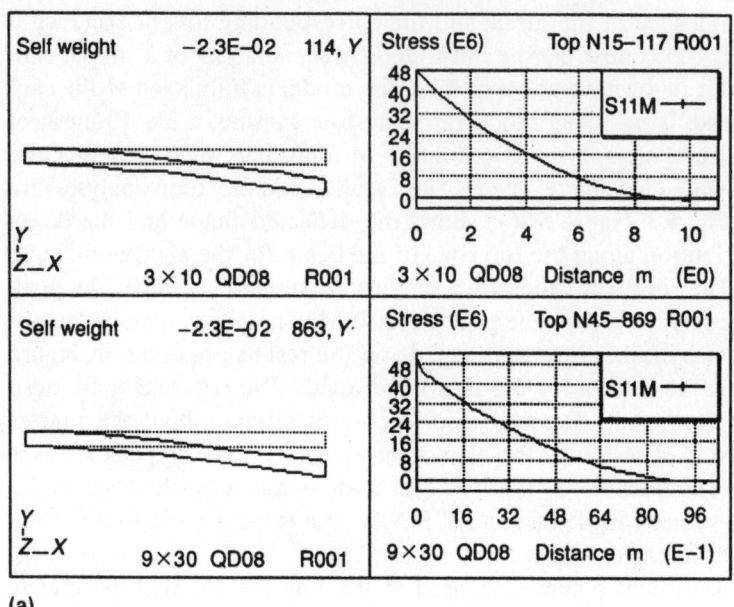

(a)

(b)

5.7 Cantilever beam results: (a) quad 8; (b) quad 4.

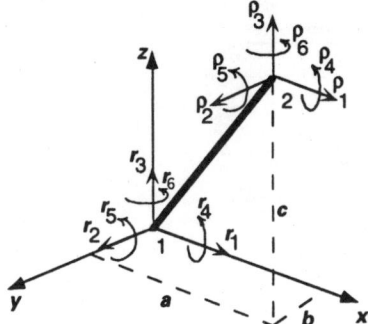

5.8 Rigid link connection.

failure can then occur by delamination. The effect of delamination has to be modelled on some occasions and special techniques are required if bending elements are being used. One such method is to represent a delamination by modelling two half-thicknesses of the plate in the area of the delamination and the full plate in non-delaminated areas. It is then necessary to join the two parts of the model and this can be done with rigid links.

Consider the rigid link shown as the thick line in Fig. 5.8. This connects nodes 1 and 2 together so that the displacements r_1 to r_6 at node 1 are related to the displacements ρ_1 to ρ_6 at node 2. Since the link is rigid the displacements r and ρ are related as:

$$
\begin{bmatrix} \rho_1 \\ \rho_2 \\ \rho_3 \\ \rho_4 \\ \rho_5 \\ \rho_6 \end{bmatrix} = \begin{bmatrix} 1 & 0 & 0 & 0 & c & -b \\ 0 & 1 & 0 & -c & 0 & a \\ 0 & 0 & 1 & b & -a & 0 \\ 0 & 0 & 0 & 1 & 0 & 0 \\ 0 & 0 & 0 & 0 & 1 & 0 \\ 0 & 0 & 0 & 0 & 0 & 1 \end{bmatrix} \begin{bmatrix} r_1 \\ r_2 \\ r_3 \\ r_4 \\ r_5 \\ r_6 \end{bmatrix}
\qquad [5.5]
$$

The above equation defines a 'multipoint constraint' that can be used to relate freedoms together. It can be used to link together two plates (or beams) so that they behave as a single entity (see also Section 4.6).

To illustrate the use of multipoint constraint (MPC) equations the example shown in Fig. 5.9 is used. This is a simply supported beam loaded by a central concentrated force, applied to the neutral surface. It has been modelled using a line of conventional beam elements. Also, it has been modelled with the central span split into two halves, with beams along the centre line of the top half, and beams along the centre line of the bottom half. These two central beams have a combined depth equal to the actual depth of the full beam. In the complete beam model all of the corresponding nodes on the top and bottom beams are connected by rigid links. This is

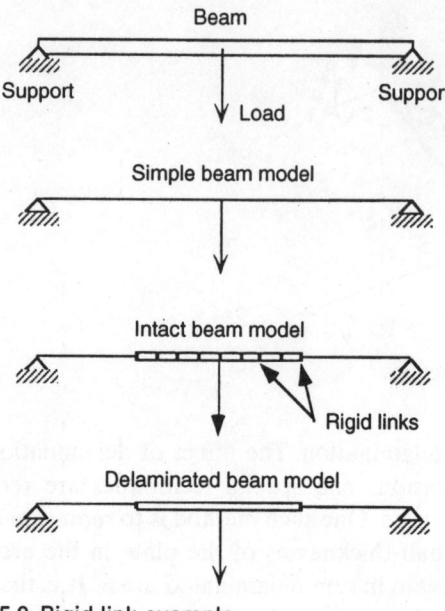

5.9 Rigid link example.

illustrated schematically by the heavy vertical lines in Fig. 5.9. Finally a beam with a delamination in the central span is modelled by only connecting the rigid links at the ends of the two central span beams. These are, of course, also connected by rigid links to the outer spans.

The displacement results for this model are shown in Fig. 5.10. The Nolink window gives the displaced shape of the simple beam model. It has a maximum deflection of 1.0×10^{-7} units at the centre of the beam. For the Link model the top and bottom central span beams have been completely joined and the displaced shape and the central deflection are identical to those of the simple beam model. The Split beam results are where the rigid links are only connected at the ends of the upper and lower half beams. In this case the lower central beam deflects more than the upper one and the maximum deflection is greater than the simple beam. This illustrates how a delamination can be modelled using beam elements. Exactly the same process, but with more links, can be used to model delaminations in plates.

5.4   Changes of mesh density

It is often necessary to increase the accuracy of the FE solution locally, typically in regions of stress concentrations or if edge effects are to be included in the analysis. As explained in Section 4.8, the accuracy of the solution can be increased by either using elements with higher-order shape functions (p-refinement), or by using a finer mesh of elements (h-refinement).

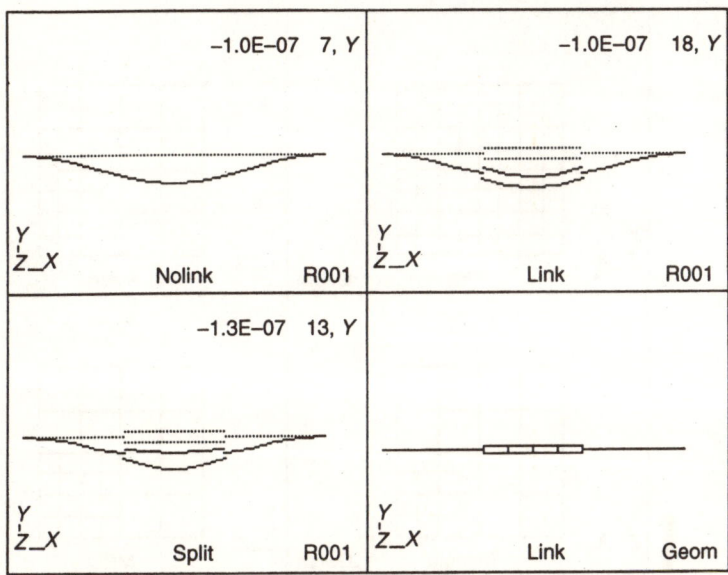

5.10 Rigid link example results.

Examples of both of these refinements are given in Fig. 5.7. They are both used in practice and generally a combination of the two is most efficient. There are some software systems that use p-refinement and automatically increase the order of the element shape functions in local regions of high stress. If h-refinement is being used then there are various ways of generating meshes for locally increasing the mesh density. Some of these are illustrated in Fig. 5.11.

The Fine mesh in Fig. 5.11 uses the same size element over the complete structure. For real structures this is generally too expensive and some form of local refinement is required. The Pave (paving) method changes the mesh density by using only quadrilateral elements. This causes some elements to become distorted with respect to the optimum square shape for the element. This technique becomes rather cumbersome unless it is done automatically by the mesh generating program. The Triangle form uses a combination of quadrilateral and triangular elements, with the triangles being used in the mesh density transition region. The MPC method has an abrupt change in the mesh density. Generally this should not be done since it will give rise to gaps along the edges of the discontinuous mesh. However, if the finite element system has the relevant facilities, compatibility can be restored by using MPCs. The constraints are defined by the shape functions of the elements in the coarser part of the mesh. This method can be very

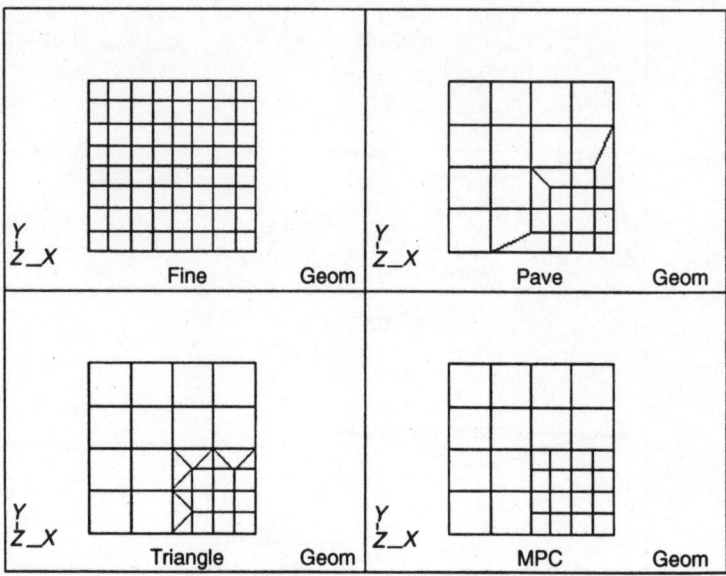

5.11 Changing mesh densities.

efficient, both for the mesh generation and for the solution process. All of the constraints can be defined and implemented at the element level.

5.5   Modelling of composite materials

A laminated composite material differs from a metal in two ways: it is a layered material built up from stacked plies of material, and, in addition, each ply is not isotropic but has directional properties with a higher stiffness in the directions of the fibres, which can change from ply to ply.

In most lay-ups the thickness is small compared with the other dimensions of the material so that it forms a plate type structure, and this is used to simplify the description. It is assumed that the strains through the thickness of the plate vary linearly with the local through-thickness, z', direction. As seen in Chapter 3, the total strain can then be written in terms of the mid-plane strain, ε^o, and the curvature, κ. Since the material properties vary from layer to layer, the stress variation through the thickness of the composite is much more complicated than that of the strains. In general there will be discontinuous changes of stress from ply to ply. This means that a simple material stiffness cannot be used for a laminated material. Instead laminate theory is employed. The stresses are integrated through the thickness of the plate. The average values of the stress give the in-plane loads **N** and the linear variation gives the couples **M**. The end loads and moments

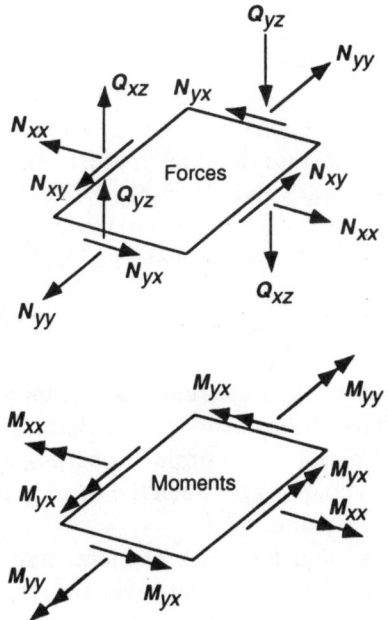

5.12 Mid-plane forces and moments.

are shown in Fig. 5.12. Using the elasticity properties of each ply, rotated to the appropriate fibre directions, the end loads and moments can be related to the mid-plane strains ε° and the curvatures κ to give the laminate stiffness properties as

$$\begin{bmatrix} N \\ M \end{bmatrix} = \begin{bmatrix} A & B \\ B & D \end{bmatrix} \begin{bmatrix} \varepsilon^\circ \\ \kappa \end{bmatrix} \qquad [5.6]$$

The full derivation of the above equation may be found in Section 3.1 (equations [3.18] and [3.19]). A are the in-plane stiffness properties, D are the bending stiffness properties and B is the coupling that arises between the bending and the membrane action. If the composite lay-up is symmetric from top to bottom then B is a null matrix and there is no coupling between membrane and bending response. All three matrices, A, B, D, are (3×3) matrices and A and D are symmetric. For a laminated composite B need not be symmetric. If extra through-thickness stiffness is introduced (z-pinning) and this is not exactly normal to the mid-plane then B is not likely to be symmetric.

For shell elements that include transverse shear strains, typically the Mindlin and the semi-Loof forms, then the effects of the transverse shear strains, γ_{xz} and γ_{yz}, must be added to the classical plate formulation. In this case the laminate stiffness matrix takes the form

$$
\begin{bmatrix} N \\ M \\ Q \end{bmatrix} = \begin{bmatrix} A & B & 0 \\ B & D & 0 \\ 0 & 0 & A_s \end{bmatrix} \begin{bmatrix} \varepsilon^0 \\ \kappa \\ \gamma \end{bmatrix}
$$

[5.7]

where A_s is the (2×2) laminate transverse material property matrix. Q is the matrix of transverse shear forces (Q_{xz} and Q_{yz} in Fig. 5.12).

The 'solid' shell has additional terms arising from the direct strains through the thickness of the shell. These special elements only have two nodes (i.e. linear shape functions) through the thickness of the element (Section 5.1). They have only translational freedoms so that both their geometrical description and their nodal freedoms correspond to standard bricks. Where they differ is in terms of their internal strain definitions. The 16- and 18-node brick elements have strains defined in the local directions of the mid-surface plane. The material stress/strain properties are then modified so that there is no Poisson coupling between the stress in the local mid-plane and strains normal to this plane.

With this idealisation it is found that the brick elements have almost exactly the same response as the corresponding plate elements. If the local coordinates in the mid-plane are ζ_1 and ζ_2 and the local coordinate normal to the plane is ζ_3 then the element interpolation functions can be written (for the 18-node brick) in the form

$$
\begin{aligned}
u' = &\left(a_0 + a_1\zeta_1 + a_2\zeta_2 + a_3\zeta_1{}^2 + a_4\zeta_1\zeta_2 + a_5\zeta_2{}^2 + a_6\zeta_1{}^2\zeta_2 \right. \\
&\left. + a_7\zeta_1\zeta_2{}^2 + a_8\zeta_1{}^2\zeta_2{}^2 \right) + \zeta_3 \left(b_0 + b_1\zeta_1 + b_2\zeta_2 + b_3\zeta_1{}^2 \right. \\
&\left. + b_4\zeta_1\zeta_2 + b_5\zeta_2{}^2 + b_6\zeta_1{}^2\zeta_2 + b_7\zeta_1\zeta_2{}^2 + b_8\zeta_1{}^2\zeta_2{}^2 \right)
\end{aligned}
$$

[5.8]

where u' is the displacement in local mid-plane coordinates. The interpolations for v' and w' are of an identical form. When these displacements are differentiated to give the strains the first line (with the 'a' coefficients) leads to the mid-plane, ε^0, strains and the second line (with the 'b' coefficients) leads to the curvature, κ. By expressing the strains in terms of mid-plane strains and curvature the material properties can be defined in terms of the **ABD** matrix. The 'solid' element includes the three in-plane strains, the two transverse shear strains and, in addition, the through-thickness strain. The equivalent **ABD** matrix then takes the form

$$
\begin{bmatrix} N \\ Z \\ M \\ Q \\ P \end{bmatrix} = \begin{bmatrix} A & 0 & B & 0 & 0 \\ 0 & C & 0 & 0 & 0 \\ B & 0 & D & 0 & 0 \\ 0 & 0 & 0 & A_s & A_c \\ 0 & 0 & 0 & A_c & A_t \end{bmatrix} \begin{bmatrix} \varepsilon^0 \\ \varepsilon_z \\ \kappa \\ \gamma \\ \gamma_z \end{bmatrix}
$$

[5.9]

Z is the through-thickness force and ε_z the corresponding through-thickness strain which are related by the material stiffness C. There are also linear variation components to the transverse shear strains which give rise to forces $\mathbf{P}$, and linear variations of the transverse shear strains, γ_z. These are coupled by the integrated material property matrix $\mathbf{A}_t$. There can also be coupling terms $\mathbf{A}_c$ but these are zero for a symmetric lay-up. Note that the zero coupling between the through-thickness stresses and strains and the in-plane stresses and strains are preserved to give accurate bending behaviour.

There are also equivalent laminate matrices representing the effects of different coefficients of expansion for the thermal loading of composite materials. These equivalent coefficient of expansion matrices are found in a similar manner to the elasticity matrices.

5.6   Examples of finite element composite analysis

To illustrate the accuracy of laminate theory, the example shown in Fig. 5.13 is used. Here a three-layer sandwich plate is modelled as a full 3D model using standard 20-node brick elements; three of these bricks are stacked through the thickness to model the laminate (right-hand windows). A plate model using laminate theory to form the material stiffness is also analysed using eight-node Mindlin shell elements (left-hand windows). The plate is loaded by a normal pressure and is simply supported around the edges. The structure has a double symmetry so that only one quarter of the plate is modelled and symmetry boundary conditions are used. The top windows show the mesh used. The deflected shape is shown in the centre windows, with the solid model having a peak central deflection of 3.5×10^{-6} units and the laminated plate 3.6×10^{-6} units. The displacements are almost identical. The lower windows show contours of the von Mises stress on the top surface of both models. The contour shapes are very similar and the stress levels are close. In this example the laminated plate model has both higher deflections and higher stresses and is probably more accurate. The composite's stresses, within the code used here, are calculated on the mid-thickness of the ply. To obtain the surface stress as accurately as possible a very thin 'witness layer' (a ply of 'zero' thickness) was added to the top and bottom surfaces of the laminate. This is quite a common modelling strategy if surface stresses are required.

The response of the nine-node Mindlin element and the 18-node 'solid' shell are compared in Fig. 5.14. Here a flat plate is cantilevered and loaded by a distributed normal pressure on the lower face of the row of tip elements. The mesh densities are identical for the two models. The conventional Mindlin plate element is called QD09 (left-hand windows) and the solid shell element is HC18 (right-hand windows). In both cases the plates

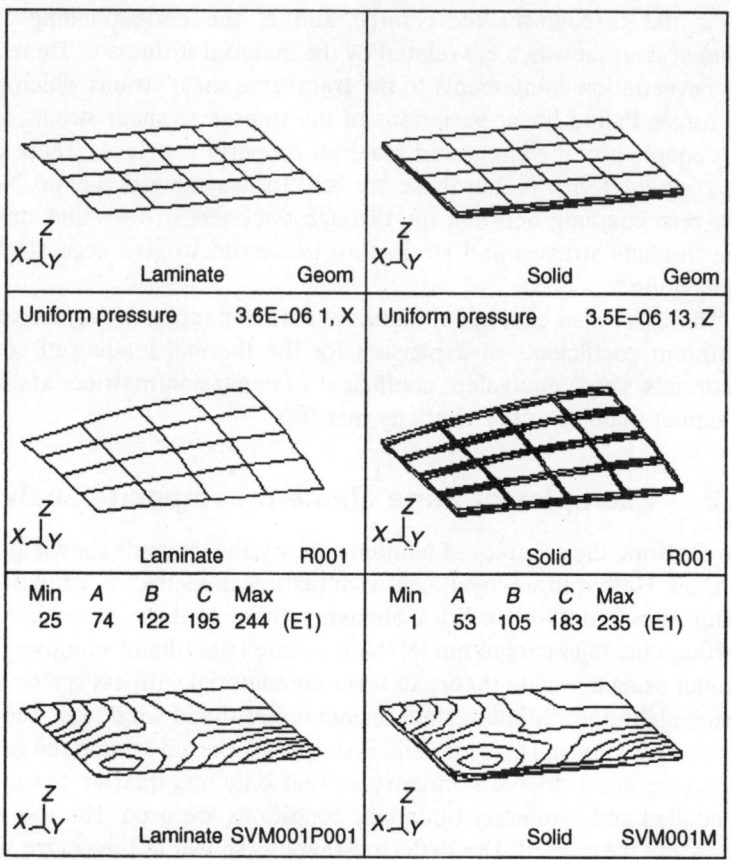

Laminate	Geom	Solid	Geom
Uniform pressure	3.6E–06 1, X	Uniform pressure	3.5E–06 13, Z
Laminate	R001	Solid	R001

Min	A	B	C	Max	Min	A	B	C	Max
25	74	122	195	244 (E1)	1	53	105	183	235 (E1)

Laminate SVM001P001	Solid SVM001M

5.13 Laminate test example.

are made of CFRP material with a quasi-isotropic lay-up of 0, 90, +45 and −45° plies. The 'solid' shell model only has one element through the thickness and uses the same laminate formulation as the shell element.

It will be seen that both models have identical maximum displacements of 2.6 units (upper windows). The distribution of von Mises equivalent stress on the top surface of both of the models is also shown (second row of windows). The stress results are identical in both the magnitude and the distribution of the contours.

As a further comparison between the Mindlin shell and the 'solid' shell the first two buckling mode shapes and buckling loads are shown, for one end loaded by an in-plane compressive force (lower two windows). It is immediately obvious that the mode shapes are identical, as are the buckling loads for each model. This comparison indicates that the large deflec-

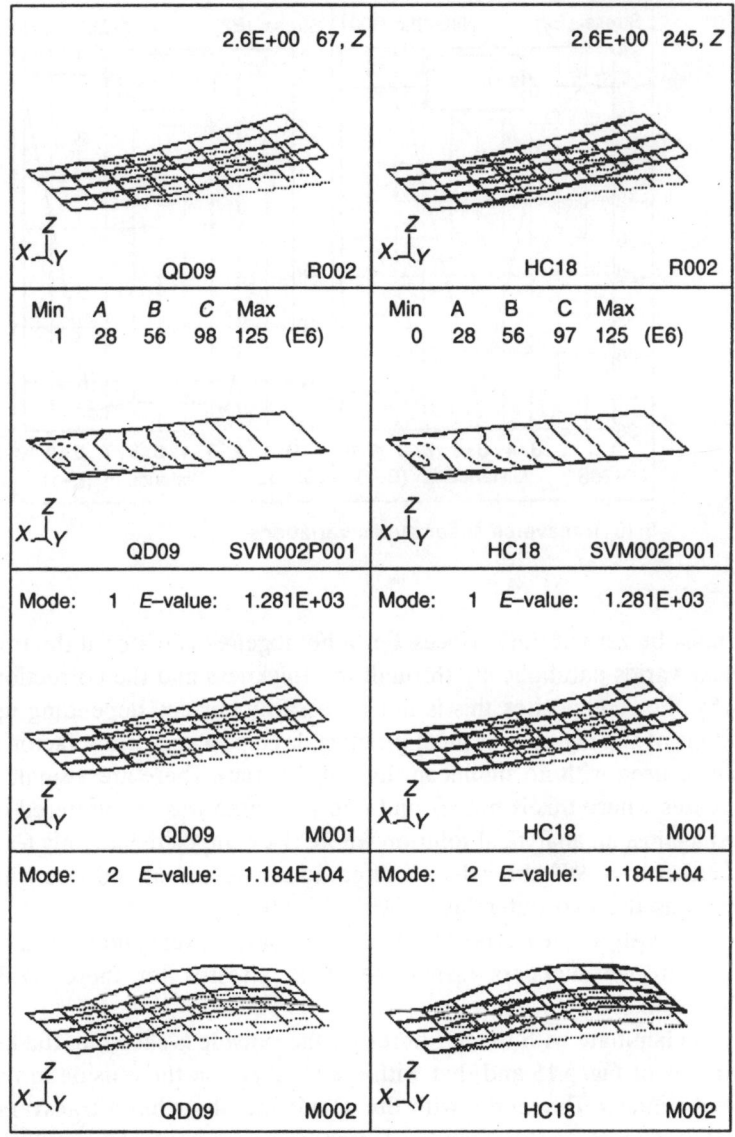

5.14 Plate and 'solid' shell comparison.

tion, geometrically non-linear, behaviour of the two elements will be the same, and this has been borne out in practice.

The Mindlin and semi-Loof elements have transverse strains, and in some codes a transverse shear factor is required to correct for the fact that the element assumes a constant transverse shear through the thickness, whereas

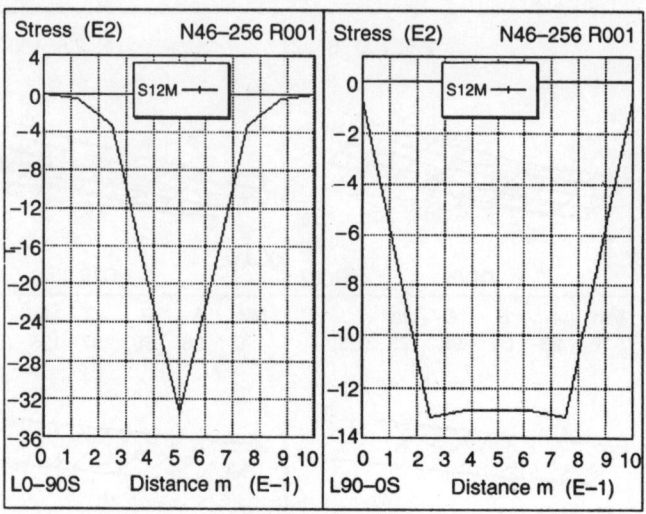

5.15 Transverse shear stress variations.

it must be zero at the surfaces. For a homogeneous material the transverse shear varies parabolically through the thickness and the correction factor is 5/6. For composites this is not necessarily correct, depending upon the details of the lay-up. For most practical lay-ups a correction factor of unity can be used with no significant loss of accuracy. There are a small number of cases where this is not so, and one such case was encountered during a test against an analytical solution where the composite had only four layers of either 0° or 90° fibres. Two configurations were analysed, one with the 0° fibres as the two outer layers with the 90° layers in the centre, and the second with this reversed; 90° as the two outer layers and 0° in the centre. The through-thickness transverse shear stresses for these models are plotted in Fig. 5.15.

The laminate with the 90° fibres on the outside is shown in the left-hand window of Fig. 5.15 and that with the 0° fibres on the outside in the right-hand window. The model with the 0° outside fibres has a transverse shear that is almost constant and a factor of unity would be applicable. The lay-up with the 90° fibres on the outside surface has transverse shear stress constant only over the inner layers and some factor very different from unity is needed.

5.7   Effect of lay-up on plate response

The lay-up of the plies can have a significant effect upon the response of a plate (as explained in Section 3.1.5). Figure 5.16 shows the displaced shapes

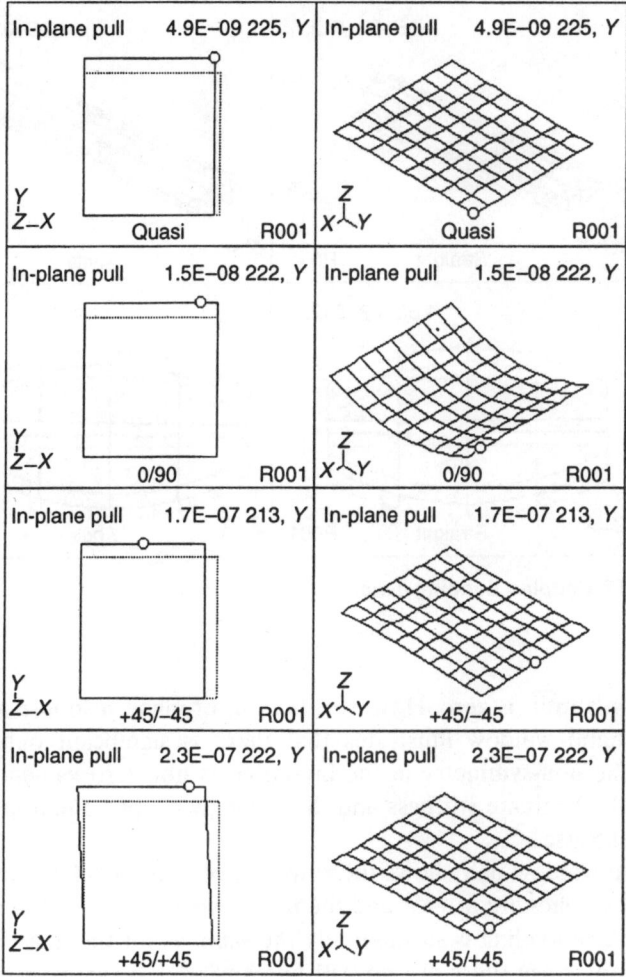

5.16 Plate lay-up variation response.

of various lay-ups. In the unloaded state all of the plates are flat and square. They are all subjected to the same in-plane tensile load. The top window shows two views of the displaced shape of a quasi-isotropic lay-up. The displacements are all in the plane of the plate. There is an extension due to the load and a small amount of lateral contraction. The response is very similar to that of a metal (isotropic) plate.

The second pair of windows shows the response of a four-ply plate, where the top two layers have fibres at 0° and the bottom two layers have fibres at 90°. In-plane there is an extension but no lateral contraction; the effec-

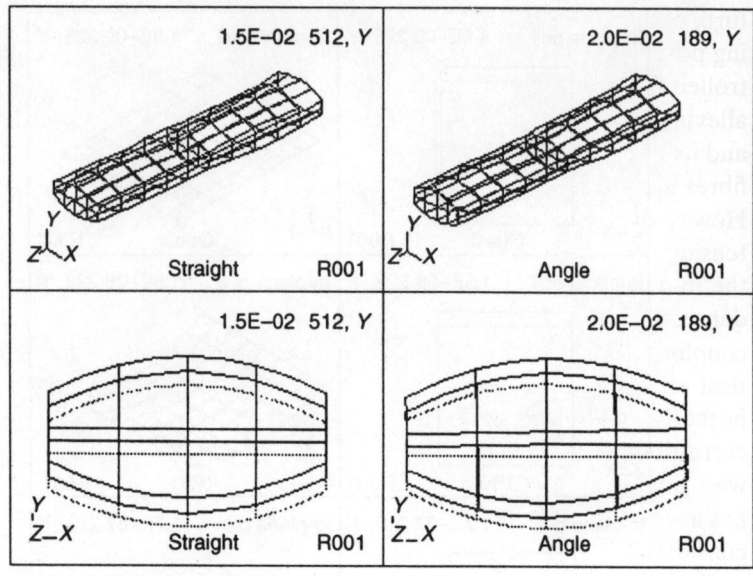

5.17 Coupled bending–twisting.

tive Poisson's ratio is zero. However, the out-of-plane response shown in the right hand window illustrates that there is significant out-of-plane bending. The non-symmetry in the lay-up gives non-zero values to the **B** matrix of the laminate stiffness and this strongly couples the in-plane and bending response.

The third window pair shows the response of a four-ply laminate where the outer two plies are at +45° and the inner two are at –45°. The response to the in-plane loading is substantially the same as for the quasi-isotropic lay-up except, since there are no stiff fibres at 0°, the extension is significantly greater than for the quasi-isotropic case. There is no significant out-of-plane response for this case even though the lay-up is not symmetric.

The bottom window pair show the response of the four-ply lay-up where all of the plies have fibres at 45°. This causes an in-plane shearing action of the plate. The lay-up is symmetric so that there is no out of plane movement. Of course, all of these responses would be predicted by standard Classical Laminate Theory.

The coupling between loadings and displacements that can arise with unbalanced or unsymmetric lay-ups can be used to advantage. One problem with aircraft behaviour is associated with aeroelastic instability, either dynamically with flutter or statically with divergence. Both effects arise as a consequence of the wing moving vertically with respect to the airflow, which causes an increase in the effective incidence of the wing which, in turn, causes more lift, making the wing move further and thereby providing positive feedback. Under some circumstances this can cause an uncontrolled increase in the lift on the wing. With a composite this effect can be alleviated by arranging for a coupling between the tension in the wing skin and its transverse shear deflection. For maximum resistance to bending, the fibres in the top and bottom skins should be oriented along the wing span. However, if they are angled slightly backwards, then the combination of tension in the bottom wing skin and compression in the top, together with the in-plane shear coupling, causes the wing to twist with a negative incidence which can cancel the induced lift due to bending. This 'aeroelastic coupling' is illustrated in Fig. 5.17. There are two models here that are identical except that the Angle model has fibres at −9° to the span direction in the top and bottom skins. The wings are loaded by an upward pressure corresponding to the lift. The Straight model (with fibres only in the spanwise direction) has no rotation of the cross-section but the Angled one has a small negative rotation. This small twist is sufficient to reduce the lift considerably.

This example serves to show that the behaviour of composite structures can be significantly different from metallic structures and this can be used to advantage. However, there are some negative consequences. It will be seen that the Angled model has a larger total deflection because the angled fibres means the wing is more flexible in bending. Also, although not shown here, there is a significant increase in the stresses in the wing skins.

5.8   Summary

Most composite structures are best modelled using plate and shell finite elements. These have to be modified to allow for laminated materials. In particular, the strains have to be expressed in terms of mid-plane strains and curvatures. Laminated composite theory can also be extended to allow for thick shell and through-thickness effects.

The finite element is approximate and the accuracy of the solution depends heavily upon the choice of elements and mesh density. The mesh should be densest where the rate of change of stress is greatest.

Special facilities within programs such as rigid links and multi-point constraints can be used to ease the mesh generation and other modelling problems.

Conventional laminate theory can be used to construct finite elements for use in the analysis of composite structures. The consequences of different lay-ups can be included directly into the FE model. The resulting coupling effects then arise naturally in the model.

Definition of composite materials in finite element analysis

6.1 Introduction

When setting up an FE model using composite materials all of the geometric modelling difficulties associated with metal structures are present and have to be addressed. In addition other modelling difficulties associated with the non-homogeneous and layered nature of the material adds extra complications and possibilities of making mistakes. These complications are discussed in this chapter.

6.2 Geometry and topology

Under this heading the needs for representing layer orientation, stacking sequence and thickness variation are considered. These pose few problems for flat or cylindrically curved laminates but any double curvature introduces severe difficulties. Firstly in identifying what the actual geometry will be (and this depends upon the method of lay-up), and secondly in providing a concise definition of that geometry in order to ease the burden of data generation. *These issues are not merely academic quibbles but are fundamental to the problem definition and will have as much bearing on the validity of the analysis as most of the mathematical and numerical approximations used in the element formulations.* A brief discussion follows on three very common geometries that couple the lay-up to the geometrical description, and many more than these three can be encountered in practice.

6.2.1 Conical surfaces

It is obvious from Fig. 6.1 that in any method of lay-up that involves parallel tape laying or broadgoods wrapping, it is impossible to achieve constant layer orientation relative to the generators of a cone in a continuous ply. Small amounts of conical taper may be accommodated by shearing of the parallel tape, but this must result in either discontinuities in fibre lay-

Conical shell

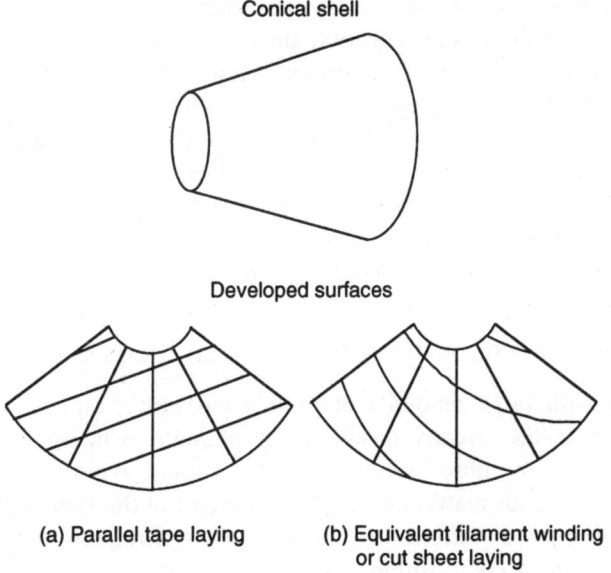

Developed surfaces

(a) Parallel tape laying

(b) Equivalent filament winding
or cut sheet laying

6.1 Conical surface lay-up.

up to match the varying layer width in the axial direction, or 'working' of the material by thinning out the wider end relative to the narrower one. If a body or surface with significant conical taper is produced, it will most probably be made by filament winding, in which case varying thickness or fibre volume fraction must always accompany constant angle winding since the same number of fibres will be present in any layer at each tapering section. The extent to which this phenomenon appears in the guise of thickness or volume fraction variation depends on the method of compaction used in the curing. Other practical solutions are to produce 3D laminates with fibre discontinuities and layers not parallel to the surface, by overlapping tape wrapping or convolute/involute preforms.

6.2.2 Double curvature in tape or sheet-laid materials

Many components are laid up from parallel tape, fabric sheet (pre-impregnated or wet laid) or broadgoods either worked over doubly curved formers or, more commonly, laid up flat and then warm worked into a pair of curved moulds. Wherever conventional lamination with continuous plies is attempted the same fundamental geometric problems occur. The thickness or the volume fraction varies with local curvature since the same amount of fibre is contained in sections of varying developed length. There is also some loss of control of the local fibre orientation as the working process causes fibre movement to accommodate the surface shape. In components cured in matched male/female moulds constant thickness and orientation of each layer relative to the natural surface coordinates would normally be assumed. The variation would then appear as a non-constant volume fraction.

It is most improbable that any FE system would permit variation of the material properties across an element and the best that could be expected is a varying thickness, and even this is not common. *Correct representation of curved shell flexural properties is thus highly improbable. At best it will be discontinuous and at worst it will be ignored altogether.*

6.2.3 Filament wound doubly curved surfaces

Components with large amounts of double curvature, especially closed shells and solids, are usually produced by filament winding. This gives precise control of the fibre orientation but it cannot avoid the universal problem of continuous materials – the containment of the same number of fibres in varying cross-sections. In spherically curved regions we may lose the nominal layered composition altogether as frequent fibre cross-over occurs so as to maintain the fibre continuity and orientation control. Large variations in thickness and/or volume fraction are inevitable and orientations will usually be a compromise between the design requirement and the mechanically achievable. *The numerical model needs to reflect what is mechanically achievable (that is, what is made) rather than an ideal design requirement.*

6.3   Fibre orientation

One obvious difference between metals and composites is the fibrous nature of the composites. The fibre direction has to be specified in the input to the FE package. Most composite structures are made by laying up sheets of material with the fibres having different directions in each sheet. In this case all of the fibres are at fixed known angles to each other, and only the direction of one layer has to be defined to specify the alignment of the lay-up. This reference direction is usually called the 'zero angle fibre', but it need not be. Associated with this is the lay-up direction defined by the manufacturing stacking sequence. This is especially important for non-balanced lay-ups since an incorrect specification of the lay-up direction will reverse the signs in the **B** matrix coupling terms, and in-plane/bending interaction will be in the wrong direction. It can also be important for stress recovery to ensure that if ply 'n' is being investigated then this is the same ply throughout the model.

In the simplest case the fibres lie at a fixed angle with respect to the global x,y axes and a single angle, θ (see Fig. 6.2), can be used. A more compli-

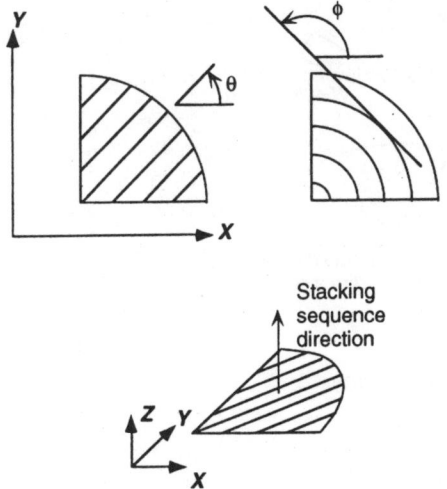

6.2 Fibre orientation.

cated lay-up can have the fibre angle changing with respect to the x,y axes, as illustrated by the angle Φ in Fig. 6.2. With some FE systems the reference fibre angle must be constant within an element. In this case the curved geometry of the fibres must be idealised by straight line segments.

For 3D problems the fibre angle definition can be complicated. In three dimensions the structure generally forms a 2D surface curving in 3D space. The fibres have to lie within this surface, but can take up an arbitrary orientation within it. Where possible the fibre angle definition should be related to the method used to construct the composite structure. If a draping program is used to define the geometry then the local fibre angle should be computed as part of the draping analysis. These angles can then be used in the FE model. If the fibre angle in the element is constant then the angle at the centroid of the element can be used. If varying fibre angles are allowed then the nodal values of the angles can be used and the element shape functions employed to interpolate for the fibre angle at the Gauss points.

If the structure is constructed by filament winding the orientation of the winding head can be used to define the fibre direction. To define this at any point the relative coordinates between the two points 'c' and 'w' (Fig. 6.3) can be used. Point 'c' is typically a node on the element and 'w' is the position of the winding head. Using the coordinates of these two points then

$$\Delta x = x_w - x_c : \Delta y = y_w - y_c : \Delta z = z_w - z_c$$

and these define a vector direction v_f of the fibre winding direction. In many circumstances these offsets are constant over large portions of the winding

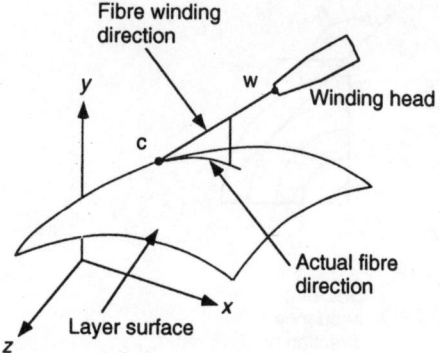

6.3 Filament wound composite.

process. The nodal coordinates of the element are known from the mesh generation process, and from these and the element shape function the vector of the normal to the surface at point 'c', v_n, can be found. Forming the cross product $(v_f \times v_n) \times v_n$ gives the local fibre direction in the plane of the layer surface. The layer stacking sequence can also be defined as being in the direction of the component of the vector 'c' to 'w' that is normal to the shell surface.

If the fibres are curving in space but remain parallel to each other then the fibre reference line method can be used. The coordinates of a general line are used to define the reference for the fibre direction and these can be given as a table of coordinates. A line is generated through these using a parametric cubic spline curve. The most convenient way of doing this is to calculate the length between points as

$$l_{i+1} = l_i + \sqrt{\left[(x_{i+1} - x_i)^2 + (y_{i+1} - y_i)^2 + (z_{i+1} - z_i)^2\right]} \quad \text{with } l_1 = 0 \quad [6.1]$$

and then use a cubic spline to interpolate for x in terms of l and, separately, for y in terms of l and for z in terms of l. The derivatives $\partial x/\partial l$, $\partial y/\partial l$ and $\partial z/\partial l$ at a point can easily be found from the cubic spline. These essentially give the fibre winding direction vector and the operations described for the filament winding procedure used to give the actual reference line direction within the surface of the element. This is necessary since, although the fibre reference line coordinate points lie in the surface, the cubic spline fit need not lie exactly in the finite element surface since these are two different approximate representations. The form of the general definition is given in Fig. 6.4. A second auxilliary line 1', 2', 3', 4', 5' can be defined to specify the normal direction from the shell surface to define the stacking sequence. This line must not pass through the shell surface, but it does not need to be

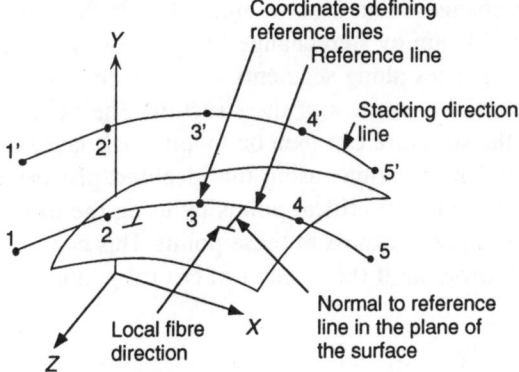

6.4 General reference line.

defined as accurately as the fibre reference line since it is only defining the direction above or below the plate.

In the worst case the fibre directions have to be specified manually. This should be avoided if at all possible since it is both time-consuming and very error-prone. This input can be done by making the fibre direction specification part of the mesh generation process. Δx, Δy and Δz can be specified at each of the mesh generation nodes and the offsets at all other points generated in the same way that the nodes themselves are generated.

6.4   Volume fraction and ply thickness variations

When the curvature of the composite skin is changing rapidly, or when a tapered filament wound component is produced, then it is likely that both the effective ply thickness and the volume fraction of the ply will be changing. If the curvature is rapid these changes will not be well controlled and their magnitude will change for each component that is made, although they are nominally identical. Such variations can probably only be treated in any realistic way by some statistical method based upon the average values. For the case of tapered filament wound structures the variations of volume fraction and ply thickness can be calculated from the geometry of the structure and the dimensions of the fibre. In principle, the individual **ABD** matrices can be re-calculated at each Gauss point of each element, in the same way that material properties can be made a function of temperature. However, this is likely to be impractical both from the point of view of computer time and from the extra storage requirements. If the taper is relatively gentle then a more practical approach is to idealise the structure so that average values of volume fraction and ply thickness are taken over areas of the surface, with abrupt changes from area to area. (This is the same idea as representing a tapered beam by step changes in the cross-section properties with constant properties along segments of the beam.) The idealised model is used to define the stiffness of the structure. The deflections and hence the strains in the structure can then be found. Basic layer stresses in the plies can be found from strains using the idealised properties. More precise stresses can be found at critical points by using the exact volume fraction and equivalent ply thickness at these points. This can be done 'by hand' outside the FE program if the number of critical points is small.

6.5   Laminate ply drop-off

One of the advantages of composite materials is that, as a part of the laying-up operations, laminae can be added or removed over relatively small areas. This allows the structure to be locally reinforced by having more layers in highly stressed areas. If plies are removed, and the ply drop-off length is smaller than the element length, then the drop-offs can be modelled by abrupt changes in material properties at element boundaries (where the element boundary is arranged to occur at the ply drop-off point). However, if the drop off is gradual, or a fine mesh has been used in the drop-off area, which is quite likely if this is a highly stressed region, then the **ABD** matrices must be changed to represent the drop-off. This can have the effect of requiring a large number of different material models (material **ABD** matrices) to be defined for the structure.

The reinforcing of a corner by using ply drop-offs is illustrated schematically in Fig. 6.5. For the vertical side the plies all drop-off so that only the first layer is left. Horizontally another form of drop-off is shown. Here the centre layers 2 and 3 are dropped-off but the two outer layers are retained and used to enclose the ends that were dropped-off. It can be very convenient to model this situation with a four-layer material model everywhere,

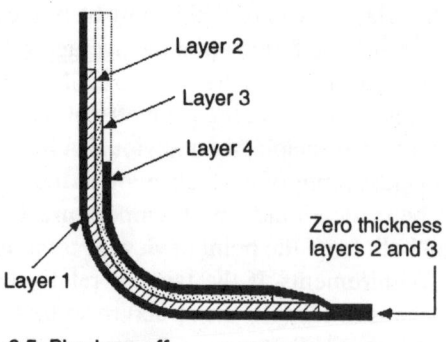

6.5 Ply drop-off.

and where a ply does not exist to give it a zero thickness when forming the **ABD** matrix. The zero thickness will mean that the layer does not contribute to the material stiffness matrix but the identity of the layer is retained. Having zero thickness also means that the layer will have no stresses associated with it. However, retaining the same number of layers throughout the skin greatly simplifies the data interpretation at the end of the analysis. Layer 4 will always be the outer layer but where it has zero thickness it will have zero stress. To draw contours of stresses on the outer layer the user has only to select layer 4. Where the thickness is zero, no contours are drawn.

This procedure can often be made considerably easier if just one lay-up is defined initially, with all four materials present with their true stiffnesses. When the mesh is generated then zero thickness for non-existent layers can be specified for the relevant groups of elements. This removes the need for the user to define many different material lay-ups. It also allows easier graphical validation of the model. The user can draw each layer separately, with the plotting routine not drawing zero thickness layers. The alternative of having many different material lay-up models is much more difficult to verify to ensure that the correct lay-up has been used at each location.

6.6   Modelling real structures

The process of modelling real composite structures can be significantly more complicated and more open to error than with metallic structures. To illustrate the difficulties the wing box form of structure shown in Fig. 6.6 is considered. This box consists of two large skin areas on the top and bottom surfaces, full depth spars at the front and rear and, in this case, a central spar. Finally, there is a series of longitudinal stiffeners. There will also be internal ribs. All of these components will have variations in the number of plies along their length, and possibly also across their width. In addition, there will be local reinforcing of the form illustrated in Fig. 6.5 at all junc-

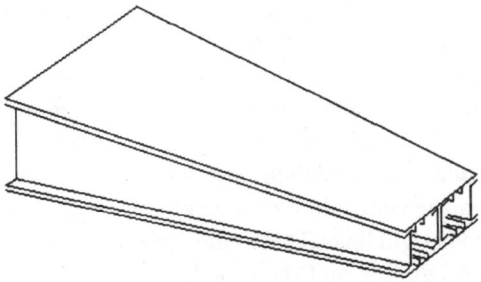

6.6 Wing box example.

tions. If a detailed analysis of the structure is required then all of these details will have to be modelled. This is significantly more complex than the equivalent metal structure where, typically, local stiffening can be represented as 'lumped area' bar members.

It makes sense to approach the modelling of the structure in the same form as it is built. The top skin can be modelled, in this case as a flat plate, but including the stiffeners attached to the plate since these are often made integral with the plate. The bottom skin is modelled similarly. If the bottom skin is a mirror image of the top skin (admittedly not a likely situation) then mirroring facilities in the mesh generation program can be used if they are available. However, care must be taken since the mirroring might be correct for defining the geometry but it may not mirror the lay-up of the composite material correctly. This can depend upon exactly how the orientation of the stacking sequence is defined and how the top and bottom surface layers are specified. However, if only a half-model of the structure is constructed, and symmetry enforced by appropriate boundary conditions, then the lay-up will be correct.

The spars are often constructed separately from the skins and, similarly, they are most easily modelled as separate components. In the manufacturing process the spars can be bonded to the skins but the modelling of the connection is not so easy. Even for co-bonded spars they are effectively separate items in many cases.

If the plate models for the skins and the spar are combined, by overlaying the horizontal portions of the spar with the corresponding skin elements, then the geometry represented will be as shown in the lower diagram of Fig. 6.7. This is not the same as the true geometry shown in the top of the figure. The offsets of the spar flanges have not been included and, to make the geometry correct, the vertical spar web is over-long. There are various ways to obtain a better model, depending upon the facilities available within the FE software being used.

One possibility is to add layers with a finite thickness but zero stiffness to the top and bottom spar flanges as shown in the lower part of Fig. 6.8. The zero stiffness layers must be wider than the skin thickness so that the geometric centre lines of the modified flanges are in the same position as the centre lines of the skins. The non-zero stiffness spar material is then correctly located and is the correct geometrical shape (upper part of Fig. 6.8).

An alternative is to model the correct skin and spar geometries and to connect them together with rigid links to allow for the offset between the centre lines. The components are modelled separately using plate elements. Corresponding nodes on the centre line of the skin elements and the spar elements are connected via rigid links. These cause plane sections to remain plane but still allow both bending and shear deflections to occur. Where it is available, especially where a large number of rigid links can be quickly

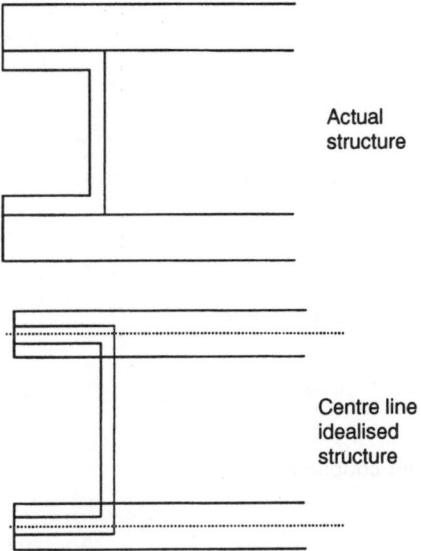

6.7 Centre line connection.

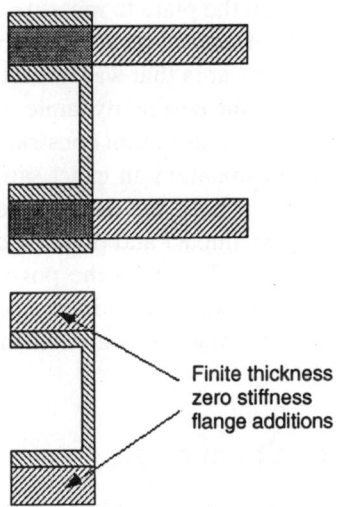

6.8 Zero stiffness spar layers.

and accurately specified, this is the most direct form coupling plate components in built-up structures. This approach is shown in Fig. 6.9.

For these situations the brick form of shell elements allows a direct representation of the true geometry, which would give the easiest way to model it. However, there are some other aspects of plate modelling where the brick form is not so convenient.

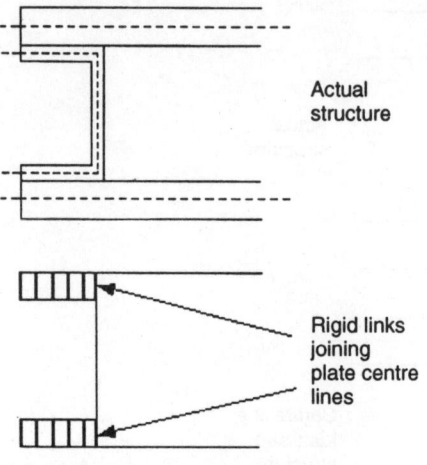

6.9 Rigid link component coupling.

Figure 6.10 shows an example of a real structure that was designed as an impact test model. It consisted of a flat plate stiffened by top hat stiffeners. Low-velocity impact tests were conducted on the plate to measure its compression after impact strength. There are various things to note about the model. There is a local mesh refinement in the area that was impacted. This degree of refinement was required to obtain the correct dynamic response. There is an abrupt change in mesh density, but multi-point constraints were applied to freedoms along this interface to maintain an exact satisfaction of compatibility. Three effects were of interest here, the dynamic response, the in-plane material degradation due to the impact and possible debonding of the foot of the stiffener from the plate. To model the possibility of debonding, layers of plates were introduced and these were coupled with rigid links. The details of the layers and the rigid links are shown in the lower two windows of Fig. 6.10.

6.7   Non-linearities in composite analysis

Non-linear behaviour can arise in various ways in composite structures. Since many composite structures consist of thin plates they are prone to large deflection non-linear response. This response is modelled in exactly the same way as for metallic structures; the composite nature of the material does not introduce any significant problems. The only difference for composites is the non-isotropic nature of the material. The orientation of the material changes with large deflections, and in almost all cases this consists of a rigid body rotation with either the stresses/strains rotated to suit the material direction (up-dated Lagrangian solution), or the material prop-

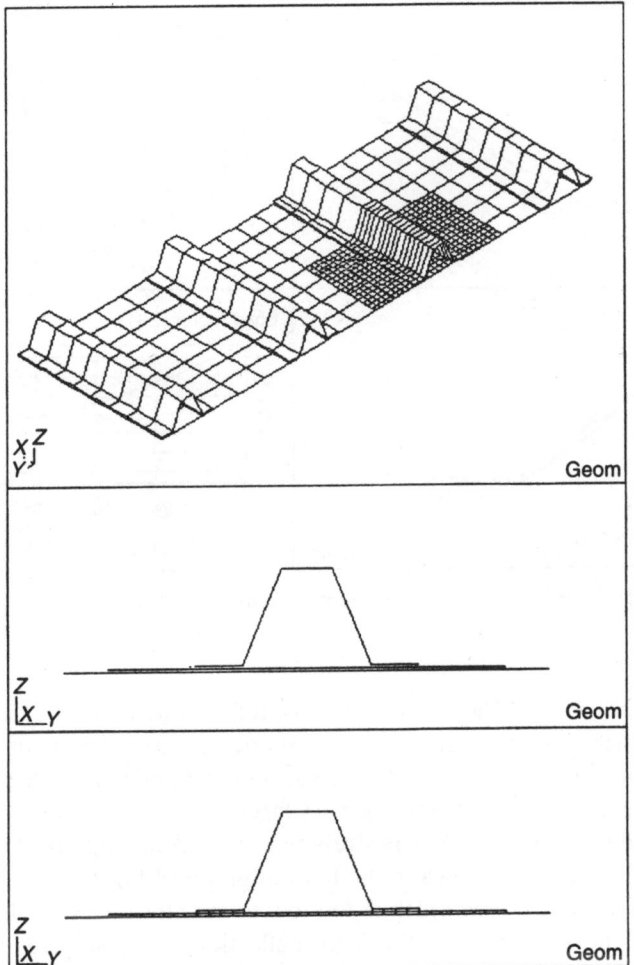

6.10 A composite stiffened panel model.

erty matrix is rotated to suit the stress/strain directions (total Lagrangian solution). The up-dated Lagrangian form is probably most suitable for these plate type problems since the stresses and strains are naturally defined in local element coordinate systems in the basic element definition.

If the structure is under a compressive load it can buckle and lose its stiffness. Fig. 6.11 shows the response of a plate under a compressive load. The top graph shows the vertical tip deflection against loading factor. As the load increases there is no significant vertical movement until the strut buckles. This occurs at a load factor of 1, the buckling load. There is then a sudden large sideways movement. This graph shows both the loading and

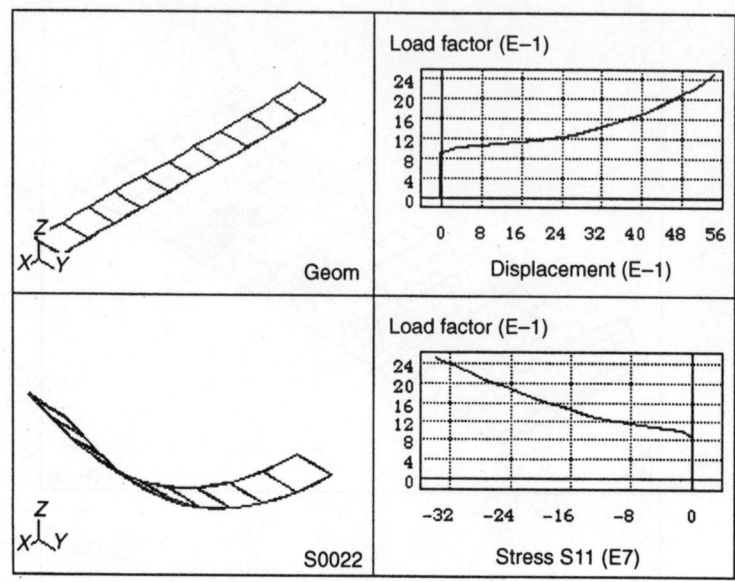

6.11 Post-buckling displacement response.

unloading behaviour. The unloading curve follows the loading exactly. The lower left window shows the deflected shape at a load factor of 2.5 (2.5 times the buckling load). This is the actual (not magnified) diplacement and it will be seen that the deflection is very large.

The loading/unloading Cauchy stress response on the top ply at the built in end of the strut are shown in the bottom graph of Fig. 6.11. There is an initial compressive stress, which is too low to register on the scale of this plot. After the strut buckles the large deflection response gives bending stresses that rise rapidly. The loading/unloading response curves are identical to each other and overlay each other on the plot.

For lower aspect ratio plate structures under compression a similar non-linear response is observed. There can be further complications for these geometries since it is common for plates to have more than one mode, with close buckling loads. In such a case the plate can buckle in an initial mode and then, after a small increase in the load factor, jump to another mode. This effect can sometimes cause difficulties in the numerical non-linear solution process.

There can be some circumstances where fibrous materials have a non-linear elasticity property. This can arise because, as the fibre composite is pulled, the inclined fibres tend to align themselves with the load and hence the stiffness increases, as shown in Fig. 6.12. Similarly, when compressed the fibres tend to become less vertical. The model in this example consists of

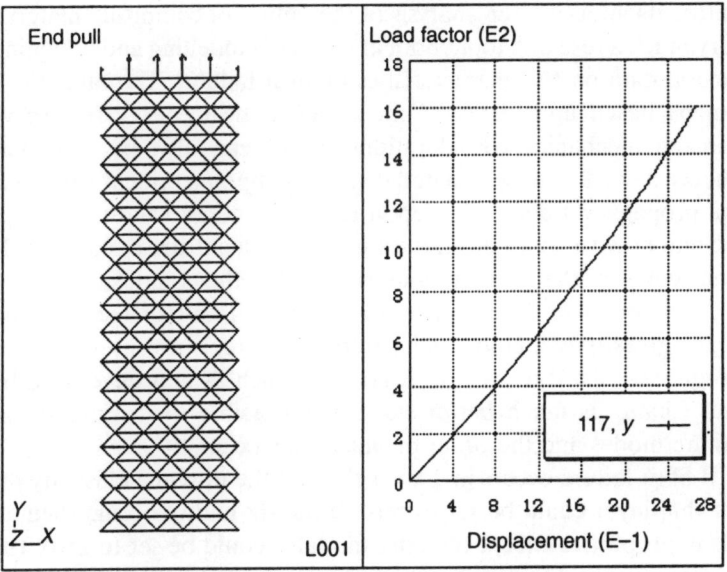

6.12 Fibre re-alignment.

diagonal and horizontal pin jointed members to represent the fibres. The model is pulled in the vertical direction. The horizontal fibres will be in compression and are likely to buckle. To represent this simply in the model they have been assigned a Young's modulus one-tenth that of the diagonal members. The load–deflection behaviour follows the same curve for both loading and unloading so that the behaviour is non-linear elastic. This effect is particularly noticeable for composites with stiff fibres and a flexible matrix, typically biological tissue material. There can also be some loss of effective stiffness in compression due to fibres buckling. This tends to give rise to a bi-linear load deflection curve with one modulus in tension and another, lower, value in compression. When the structure is under a tensile load it becomes stiffer and when the load is compressive it is more flexible. The tensile behaviour only is shown in Fig. 6.12.

6.8   Failure models

Most structural composite materials are brittle and this can lead to a non-linear behaviour involving brittle fracture. There are three main forms of failure: fibre fracture, matrix cracking (see Section 3.2) and delamination. All composites will have some degree of delamination, broken fibres and cracked matrix, but this is only significant if the damage grows to occupy large and/or critical areas of the material so that the component is starting to lose its integrity. The analysis of the failure of composite materials is still very much a research topic, both in terms of modelling and obtaining experimental data on the characteristics of such failures. The numerical models can be based upon stress/strain values, critical energy release values or damage mechanics considerations. There can also be combinations of effects; a crack can be initiated depending upon a critical stress value, with the propagation depending upon released energy values. Other problems include how to combine stresses in a multi-axial stress states, and interactions between the different materials within the composite.

An interactive failure criterion, such as Tsai-Hill (see Section 3.3.2), which gives only a global indication of failure, is not useful for modelling progressive damage growth. A criterion such as that developed by Chang and Chang[1] is much better since it is possible to distinguish individual failure modes and the order in which they occur.

If fibre failure occurs in a layer then all the in-plane elasticity properties for that layer could be set to zero. If matrix failure occurs then all the in-plane properties except the fibre modulus could be set to zero. This would be done at the Gauss integration points. In fact, in order to preserve numerical stability of the solution, the properties would be set to a value of 0.001 times their intact stiffness, rather than to zero. This still makes their effective stiffness zero but they retain a 'numerical' stiffness. A new **ABD** matrix is then formed for this section using the degraded properties. Such a model would not include delamination failure.

There are variations to the latter approach which would represent the material more accurately. For example, woven material does not lose its strength entirely after initial failure, a significant stiffness being retained. Associated models do not degrade the stiffness property to zero immediately, but do so over a stress range.

If there is an existing crack in the material then fracture mechanics can be used to decide if the crack will propagate (see Chapter 8). The energy released, ΔG, by the crack advancing over an area ΔA is found and the ratio $\Delta G/\Delta A$ is compared with the critical energy release rate, $\Delta G_c/\Delta A$, for the material. If the critical energy is exceeded then the crack is assumed to propagate. There are three possible modes of crack propagation, shown in Fig. 6.13, and each has different critical energies. One as yet unresolved problem is how to combine these energies if mixed mode crack propagation occurs. In all current fracture mechanics models the analyst has to model the possible crack surfaces within the FE model. This can be done in various ways. One possibility is to use a rigid link model, as shown in Fig. 5.9. The plate is modelled as two (or more) parallel plates joined together with rigid links. The forces within the links are used to calculate the internal forces and hence the energy release as the crack opens. If the

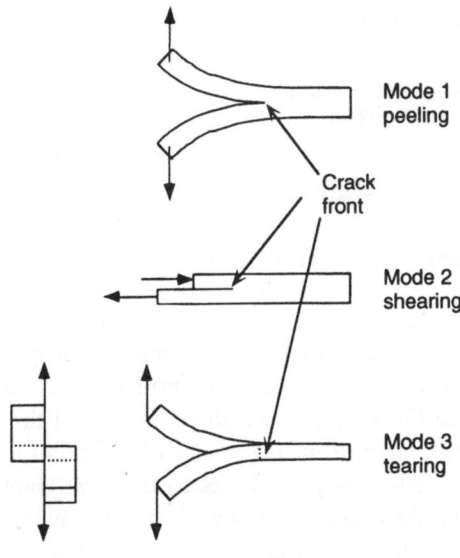

Mode 1
peeling

Crack
front

Mode 2
shearing

Mode 3
tearing

6.13 Crack opening modes.

critical energy is exceeded then the corresponding link can be 'broken' to simulate the crack growth.

An alternative is to employ a special interface element in place of the links, using the concept of damage mechanics. The element, which has the load–deflection material properties shown in Fig. 6.14,[2] models both the initial fracture and the subsequent crack growth. The element is sometimes referred to as a 'resin-rich' element. In practice, for realistic values of modulus, fracture load and critical strain energy release, a fine mesh of these elements is required in the region of the crack tip for accurate results. Also, the crack propagation phase has a negative stiffness and this requires special care in the choice of solution method. The element includes the effect of first failure with the fracture load F_c, and the subsequent fracture mechanics is incorporated in the choice of the area under the load–deflection curve. The latter is made equal to the critical energy release. For a single mode of crack opening this approach can work well. However, multimode crack propagation is not straightforward. Only one area is available in the model but there are three critical energy release modes, which requires an approximation to be made.

The interface element is still being investigated and various computational problems remain to be solved. Apart from the approximation required for multimode cracks, a fine mesh is required at the crack tip in

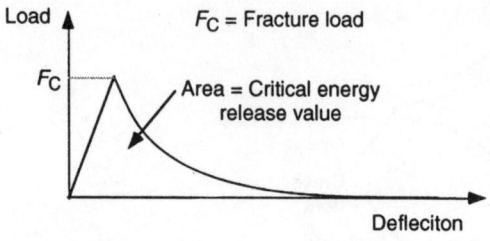

6.14 Fracture element load/deflection curve.

order for the crack to propagate smoothly. For real material properties this requires a *very* fine mesh. If the mesh is not fine enough the crack propagates to one Gauss point in the element and then stops until the stresses from the crack tip singularity have risen sufficiently at this point to exceed the fracture load, and the crack then propagates to the next Gauss point. In practice, to make the solution viable, an automatic mesh refinement algorithm is required to move a fine mesh along with the crack front. This procedure becomes very cumbersome and difficult to do for a surface crack propagating into a solid. It is almost impossible to do for multiple cracks that can merge, and for handling the crack propagation across structural discontinuities, e.g. under the foot of a stiffener.

6.9   Determination of energy release rates

One requirement for crack propagation modelling is the necessity to determine whether (and how) a crack will propagate. If a sharp crack or delamination exists in a structure then there will be a singular stress field around the crack tip. If the material failure is determined solely on a strength basis then this singularity would exceed the strength and the crack front would travel through the structure, advancing at the speed of sound within the material. Such behaviour is only observed for very brittle materials. For most materials the crack might not propagate and, when it does, it travels at a speed much lower than the speed of sound. This has led to other theories being developed to explain crack propagation, notably linear elastic fracture mechanics (see Chapter 8). The latter assumes that around the crack tip there is a very localised area which, for metals, yields and so limits the peak stresses. As already mentioned, the crack will then grow if the energy it releases by propagating is greater than a critical energy release rate for the material. If the energy caused by propagation is less than the critical value the crack will not propagate. The critical energy release rate, G_c, is a measured material property.

There are various methods for calculating the energy release rate within a theoretical model. Classical methods are typified by the J-integral method,

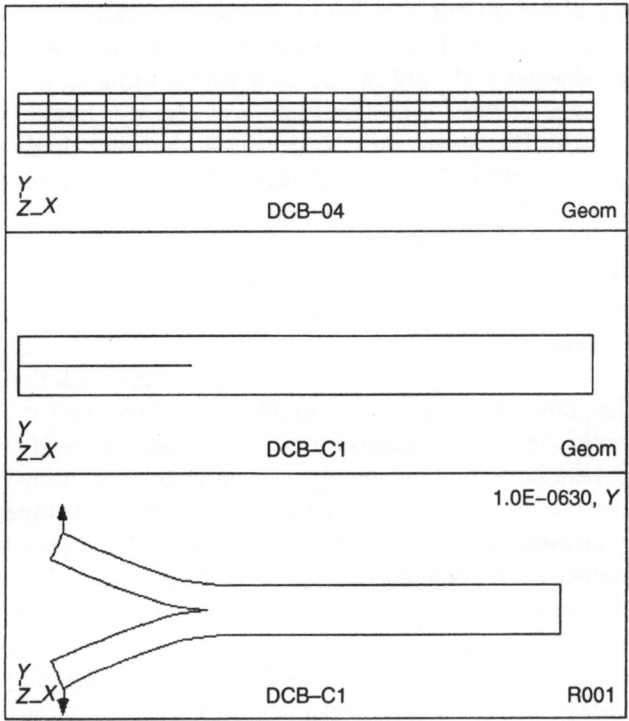

6.15 DCB model for Virtual Crack Closure.

which is not very suitable for finite element analysis and, instead, other techniques such as the Virtual Crack Closure (VCC) method are used.

A typical test specimen for determining G_c, the double cantilever beam (DCB), is shown in Fig. 6.15. This is a cantilever where a crack has been induced at the free end, in this case on the centre line of the beam, although it could be off-centre. The model is loaded by applying equal and opposite displacements at the free end, as shown by the arrows in the bottom window. The displacements at the nodes on the crack surface adjacent to the tip are recorded. Forces are then applied at these nodes of a sufficient magnitude just to join the nodes together, that is to close the crack by one node spacing. The work required to close the crack is then the forces times the displacements before the crack was closed. This work would be equal to the energy released if the crack opened by one node. If the energy released per unit area is greater than the critical value, G_c, then the crack will propagate. In practice, provided the crack tip is not near to either end of the beam, the local stress field around the tip will be the same when the crack opens by one element; that is the two stress fields are self-similar. In this case only one calculation need be done, as follows. The displacements are recorded as before but now the inter-element forces at the crack tip node are found. These will be identical to the forces required to close the crack at the adjacent node and can be used instead of these.

One advantage of this form of calculation is that it is based on energy rather than stress. This means that the stresses at the crack tip do not need to be found accurately and a very fine mesh at the crack tip is not required. However, the mesh must be sufficiently fine for the stress field to be self-similar and for the implied use of superposition to be valid.

The problem is complicated by real cracks in that, as shown in Fig. 6.13, there are three modes of crack opening and they all have very different critical energy release rates. If a pure crack opening mode is present then the appropriate G_c is used. Typically, for the DCB specimen with the crack on the centre line this is a pure mode I crack. However, if the crack is off-centre then the crack will involve a mixture of mode I and mode II. The VCC method can be used to find the energy in each mode by splitting the stress fields into symmetric and anti-symmetric components with respect to the crack line. The symmetric components give the mode I energy release and the anti-symmetric components the mode II. In practice this reduces to taking appropriate combinations of forces and displacements for calculating the energy release. There are other methods of calculating the mode components of the energy release and, although they all agree to within a small percentage with the total energy released, they can differ in their estimates of the mode components. It is not clear yet which calculation is most relevant. The whole process is yet more complicated by the fact that there is no well-established way of combining the individual mode energy releases to give a single equivalent energy release. Such an equivalent value is needed since there is assumed to be only one crack front, even when there are multiple modes involved.

The single step VCC method described above is a commonly available technique but it does have limits to its applicability. A more general method is the two-step crack opening method. Here, a first run is conducted and the inter-element forces on the faces of the elements at the crack tip are recorded. The crack is then advanced across these elements and the new model is re-run to find the displacements across the newly opened elements. These displacements can then be multiplied by the previously recorded forces holding the elements together to give the energy released. This approach does not require the stress field to be self-similar between opening and closing.

6.10 Crack propagation modelling

A number of crack propagation models are being developed but these are all still at the research stage (see also Chapter 9). One method is to use the interface element previously mentioned. This element has the advantage that it includes both a failure load, to give a stress-based crack initiation model, and an energy release component for fracture mechanics (propagation). There are problems though. The negative stiffness associated with the energy release can be numerically unstable and requires special solution methods. Also, for realistic values of failure strength and energy release, a fine mesh is required near the crack tip. In practice this means that some form of moving mesh algorithm is required as the crack propagates, in order to have realistic model sizes (see Section 4.8).

Another approach has recently been developed by workers in Sweden.[3,4] The structure is loaded incrementally and the crack allowed to grow by a small amount defined by the load energy release rates. The model also allows for large deflection, and buckling of the structure, either locally associated with the delamination, or globally. It also includes a contact model to permit the two surfaces of the crack to come together in some areas.

For energy release fracture mechanics there must be an existing crack, or we must use a stress model for crack initiation. For thin shells we can use the same rigid link model as for stress-based failure, but calculate link forces only. The virtual crack closure method is used to determine the energy release rates. An equivalent failure energy of the form

$$\gamma_j = \left[\frac{G_I[X_j P]}{G_{Ic}}\right]^a + \left[\frac{G_{II}[X_j P]}{G_{IIc}}\right]^b + \left[\frac{G_{III}[X_j P]}{G_{IIIc}}\right]^c \quad j=1,2,\dots,n \quad [6.2]$$

is defined where:

X_j = the position variables,
P = the loading,
G = the energy release rates,
a,b,c = material constants,
n = the number of nodes on crack front,

The position variables are chosen to satisfy the criterion

$$\gamma_j(X_j) - 1 = 0 \quad j=1,2,\dots,n$$

The crack front is then moved to the position defined by the X_j values. Again, in practice, this requires a fine mesh region around the crack front and some form of moving mesh algorithm is required to model any real problem. As with the interface element, this procedure becomes computationally very complicated for real geometries.

6.11 Micromechanical models using finite element analysis

The FE method can also be used to model the behaviour of the material on the micromechanical level, either for a single layer or for a group of layers.

Typically a small section of a plate is considered, say a square in plan view, where the side lengths are equal to the plate thickness and a complete depth of lay-up. This section is then modelled in detail using brick elements to represent the matrix, typically assuming isotropic properties for this, and thin membrane plates to represent the fibres. These plates will have anisotropic properties. The plates are positioned corresponding to the centre line of the fibres. This model is small and a fine mesh of elements can be used.

The model can be used for various purposes but one thing that can be done is to load it as if it were in a testing machine. All the through-thickness edges can be held fixed and then boundary displacements applied to give unit strains to the model. For the first load case a set of displacements is applied to opposite faces so that the only strain is in the x-direction, and this has a unit value. This is displacement set D_1. The reaction forces on all of the fixed boundaries are computed and these give a set of reaction forces R_1. This process is then repeated, but now applying a set of displacements D_2 that give rise to a unit y-strain. Again the reaction forces are computed and saved as R_2. This is repeated for all of the other possible strain components, typically the in-plane shear, unit curvature κ_{xx}, unit curvature κ_{yy}, unit twisting κ_{xy}, unit transverse shear ε_{xz} and unit transverse shear ε_{yz}. All of these displacements are combined into a single matrix:

$$\mathbf{D} = [\mathbf{D}_1 \mathbf{D}_2 \mathbf{D}_3 \ldots] \qquad [6.3]$$

and the boundary reaction forces into the single reaction matrix

$$\mathbf{R} = [\mathbf{R}_1 \mathbf{R}_2 \mathbf{R}_3 \ldots] \qquad [6.4]$$

The equivalent material property matrix is then given by:

$$\mathbf{E} = \frac{\mathbf{R}^t \mathbf{D}}{V} \qquad [6.5]$$

where V is the volume of the model.

A result for a 30% fibre volume unidirectional carbon fibre/epoxy layer, compared with the Halpin–Tsai equations[5] (H-T), yields the comparison:

FE: $E_{11} = 78.5\,\text{GPa}$, $E_{22} = 7.78\,\text{GPa}$, $G_{12} = 2.08\,\text{GPa}$, $v_{12} = 0.35$

H-T: $E_{11} = 78.5\,\text{GPa}$, $E_{22} = 2.96\,\text{GPa}$, $G_{12} = 1.10\,\text{GPa}$, $v_{12} = 0.35$

It is seen that the results for 'longitudinal' properties (E_{11} and v_{12}) compare well, but there are big discrepancies between the 'transverse' properties (E_{22} and G_{12}). The details of the modelling, by whichever approach, are clearly far more critical for the latter properties.

These material properties can then be used as the **ABD** matrix in a shell analysis. From this the mid-plane strains and curvatures at a point can be

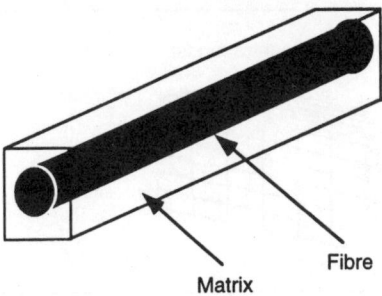

Fibre

Matrix

6.16 Micro-mechanics fibre/matrix model.

computed as usual. However, a more detailed stress recovery can be made by applying these strains to the 'test specimen' model. In fact, the stresses for the unit strain load cases will be available from the preliminary material property calculation, so that these can be linearly combined using the actual strains at a point as multiplying factors to give a highly detailed stress distribution calculation. If the 'test specimen' mesh is fine then components such as interlaminar shear and through-thickness direct stresses will also be recovered. Also, other runs of the 'test specimen' (possibly another model of larger dimensions) can be used to investigate free edge stresses and associated stress concentrations.

The above approach can be used for other problems where highly detailed real geometry effects can be 'smeared' to give an average overall behaviour. The accuracy of the approach then depends upon the micromechanical model used. A typical case is the single fibre/matrix structure shown in Fig. 6.16. The structure can include the fibre material, the matrix material and the interface between fibre and matrix. This combination is then modelled using the fine mesh shown in Fig. 6.17. The six unit strain boundary conditions, to get the equivalent in-plane stiffness properties, are applied as imposed displacements on all boundary nodes.

For example, for a constant unit shear strain, ε_{yz}:

$$\varepsilon_{yz} = 1 = \frac{\partial v}{\partial z} + \frac{\partial w}{\partial y} \qquad [6.6]$$

and this is satisfied if the displacements $v = 0.5z^*$ and $w = 0.5y^*$ are applied on the boundary nodes, z^* and y^* being the coordinate values for the boundary nodes. The other five unit strains are dealt with in a similar manner.

Such models also give the detailed stress distributions around the fibre once the actual strains have been determined. The equivalent smeared properties are used in a global model (exactly equivalent to the laminate

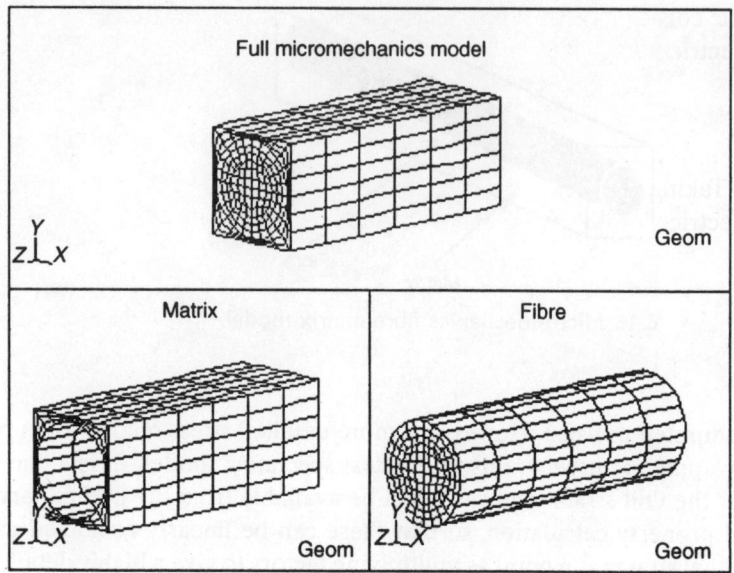

6.17 Mesh for micromechanics model.

ABD material matrix), the problem is solved and the strains at any point computed. The stresses from the unit strain load cases are then multiplied by the appropriate actual strain, and summed to give the detailed stress field around the individual fibre at this point in the material.

6.12 Smart composites: the inclusion of piezo-electric effects

There is currently much interest in producing 'smart' composites where, for example, areas of piezo-electric materials are included in the lay-up. A piezo-electric material is one where the elastic stress/strain behaviour couples with the electric voltage charge behaviour. In one dimension the elastic equilibrium equation (ignoring body forces) becomes

$$EA\frac{\partial^2 u}{\partial x^2} - FA\frac{\partial^2 V}{\partial x^2} = 0 \qquad [6.7]$$

and the electric equilibrium equation is

$$FA\frac{\partial^2 u}{\partial x^2} + GA\frac{\partial^2 V}{\partial x^2} = 0 \qquad [6.8]$$

where u is the displacement at a point, V is the voltage and A the area of the element. The material properties are E, Young's modulus; G, the dielectric constant at constant strain; and F, the piezo-electric constant. The electric field is

$$e = -\frac{\partial V}{\partial x} \qquad [6.9]$$

Taking the basic structural variables as the displacements u, v, w, and the electric variable as the voltage V, then the FE field interpolation is

$$\begin{bmatrix} u \\ v \\ w \\ V \end{bmatrix} = \mathbf{u} = \mathbf{Nd}$$

$$= \begin{bmatrix} N_1 & 0 & 0 & 0 & N_2 & 0 & 0 & 0 & \ldots \\ 0 & N_1 & 0 & 0 & 0 & N_2 & 0 & 0 & \ldots \\ 0 & 0 & N_1 & 0 & 0 & 0 & N_2 & 0 & \ldots \\ 0 & 0 & 0 & N_1 & 0 & 0 & 0 & N_2 & \ldots \end{bmatrix} \begin{bmatrix} d_1 \\ d_2 \\ d_3 \\ d_4 \\ \cdot \\ \cdot \\ \cdot \\ d_n \end{bmatrix} \qquad [6.10]$$

where $\mathbf{N_i}$ is the usual shape function at node i and d_1, d_2 and d_3 are the displacements at node 1 and d_4 is the voltage at node 1, etc. Equation [6.10] can be compared directly with equation [4.24]. The mechanical and electrical strains can be written in a single matrix equation

$$\varepsilon = \mathbf{D^t u} = \mathbf{D^t Nd} = \mathbf{Bd} \qquad [6.11]$$

where $\mathbf{D^t}$ is the matrix of differential operators including the the usual displacements of derivatives for the mechanical strains, and also the voltage derivatives for the electrical field. Equation [6.11] compares with equation [4.25].

The combined elastic/piezo-electric stiffness form of the equilibrium equations is then

$$\mathbf{P} = \int_V \mathbf{B^t EBd}V\mathbf{d} = \mathbf{kd} \qquad [6.12]$$

where the material property matrix, $\mathbf{E}$, is now

$$\mathbf{E} = \begin{bmatrix} \mathbf{E} & \mathbf{F} \\ \mathbf{F^t} & -G \end{bmatrix} \qquad [6.13]$$

The property matrix $\mathbf{E}$ is symmetric but the minus sign on the dielectric constant variable, G, means that it is not positive definite. It then follows that neither the element stiffness matrix nor the assembled stiffness matrix is positive definite. The matrix solver must be able to cope with non-positive definite matrices.

The nodal displacement vector now has displacements and a voltage as nodal freedoms. The nodal load vector now has terms for forces corresponding to the nodal displacements and charges corresponding to the voltage. The piezo-electric coupling term, **F**, couples the elastic and electric fields together so that an applied voltage or charge can cause a displacement or a reaction force. Similarly, an applied force or displacement causes a voltage or charge change. This coupling allows the structural response to be modified by an applied voltage to represent one area of piezo-electric material acting as a sensor and another area acting as a force transducer.

Layers of piezo-electric film can be included in a composite lay-up to form a piezo-electric composite material. Laminate theory can be applied to this to form an equivalent laminate **ABD** matrix including the elastic, the electric and the piezo-electric effects. Provided sufficient forces can be generated by the piezo-electric material and, provided it can withstand the resulting stresses, a very versatile structure can be produced. However, many current piezo-electric materials do not generate much force and they are very brittle and weak. It is worth noting that piezo-electric materials have been used for some time. The crystals in crystal-controlled oscillators are piezo-electric and it is the low loss factor associated with mechanical vibrations that make them such stable and precisely defined frequency sources.

The dynamic behaviour of piezo-electric materials also presents some numerical difficulties. In addition to the non-positive definite form of the stiffness matrix there is no inertia associated with the electric variables. This leads to the possibility of some modes having infinite frequencies. However, if an eigenvalue extraction scheme such as the Lanczos method is used, these modes are not computed and present no numerical difficulties.

6.13 Summary

The description of real composite structures can be difficult to model. Fibre directions for materials where the fibres are arbitrarily oriented in 3D space are difficult to define. Variations of lay-up thickness and volume fraction have to be approximated in the model. These parameters have to be constant over an element in most systems.

Ply drop-offs can be modelled by assigning zero material properties to the dropped-off plies. For real structures either rigid links or fictitious plies can then be used to model the offsets between plate and shell centre-line surfaces.

The thin plate-like nature of composite structures means that they are likely to be prone to vibration, buckling and large deflection behaviour. These can all be modelled using standard FE theory.

Composite materials can fail by matrix and fibre fracture (material degradation) and by ply delamination (fracture mechanics). Models for material degradation are well developed but the modelling methods for delamination growth are still being developed.

Detailed 3D FE models can be used to construct micro-mechanical models of either individual fibres or bundles of fibres.

6.14   References

1 Chang F K & Chang K Y, 'A progressive damage model for laminated composites containing stress concentrations', *J Compos Mater*, 1987 **21**(9) 834–855.

2 Chen J, Crisfield M A, Kinloch A J, Matthews F L, Busso E & Qiu Y, 'Predicting progressive delamination of composite material specimens via interface elements', *Mechanics of Composite Materials & Structures*, 1999 **6**(1) 1–17.

3 Singh S, Asp L E, Nilsson K-F & Alpman J E, *Development of a model for delamination buckling and growth in stiffened composite structures*, TN 1998-53, Bromma, The Aeronautical Research Institute of Sweden (FFA), 1998.

4 Nilsson K-F, Asp L E, Alpman J E & Nystedt L, *Delamination buckling and growth for delaminations at different depths in a slender composite panel*, TN 1999-04, Bromma, The Aeronautical Research Institute of Sweden (FFA), 1999.

5 Halpin J C, *Primer on composite materials: Analysis*, Lancaster, Penn, USA, Technomic Publishing Co, 1984.

Part IV
Analytical and numerical modelling

C. SOUTIS

This final part of the book is sub-divided into several chapters, each dealing with an important issue in the application of FRP composites. The approach is to present analytical models alongside FE representations. The former are always needed, when available, to confirm the FE models. Specific issues addressed include: inter-laminar stresses, fracture, delamination, joints and fatigue.

Interlaminar stresses and free-edge effects

7.1 Introduction

The classical laminate theory (CLT) discussed in Section 3.1 provides a simple and direct procedure for calculating stresses and strains. It is, however, not very accurate since it does not satisfy equations of elasticity at every point. This can be shown through a simple example. Consider a laminate subjected to uniaxial tensile loading in the x-direction ($N_x = N_0$, $N_y = N_{xy} = 0$), as shown in Fig. 7.1. The mid-plane strains and plate curvatures can be obtained from equation [7.1] (see equations [3.18] and [3.19]) and then the lamina stresses calculated according to equation [7.2] (see equation [3.15]). In general, the lamina stresses σ_y and τ_{xy} will not vanish even on a free boundary parallel to the x-axis; for example at the edge of the laminate $y = \pm b$ as shown in Fig. 7.1. This violates the equilibrium boundary conditions on the edge.

$$\begin{bmatrix} N \\ M \end{bmatrix} = \begin{bmatrix} A & B \\ B & D \end{bmatrix} \begin{bmatrix} \varepsilon^0 \\ \kappa \end{bmatrix} \tag{7.1}$$

$$\begin{bmatrix} \sigma_x \\ \sigma_y \\ \tau_{xy} \end{bmatrix}_j = \begin{bmatrix} \overline{Q}_{11} & \overline{Q}_{12} & \overline{Q}_{13} \\ \overline{Q}_{12} & \overline{Q}_{22} & \overline{Q}_{23} \\ \overline{Q}_{13} & \overline{Q}_{23} & \overline{Q}_{33} \end{bmatrix}_j \begin{bmatrix} \varepsilon_x^0 \\ \varepsilon_y^0 \\ \gamma_{xy}^0 \end{bmatrix} + z \begin{bmatrix} \overline{Q}_{11} & \overline{Q}_{12} & \overline{Q}_{13} \\ \overline{Q}_{12} & \overline{Q}_{22} & \overline{Q}_{23} \\ \overline{Q}_{13} & \overline{Q}_{23} & \overline{Q}_{33} \end{bmatrix}_j \begin{bmatrix} \kappa_x \\ \kappa_y \\ \kappa_{xy} \end{bmatrix} \tag{7.2}$$

Also, CLT ignores shear deformations of layers because of the assumption that the bond between two laminae is non-shear deformable. However, when the generalised plane stresses are applied to a laminate, different layers tend to slide over each other because of the differences in their elastic constants. The two most important properties for determining the presence and magnitude of interlaminar stresses are Poisson's ratio, υ_{12}, and the shear coupling term, η_{12}.

Since the layers are elastically connected through their faces, shear stresses are developed on the faces of each layer. The transverse stresses (σ_z, τ_{xz}, τ_{yz}) thus produced (see Fig. 7.1) are negligible in the regions away

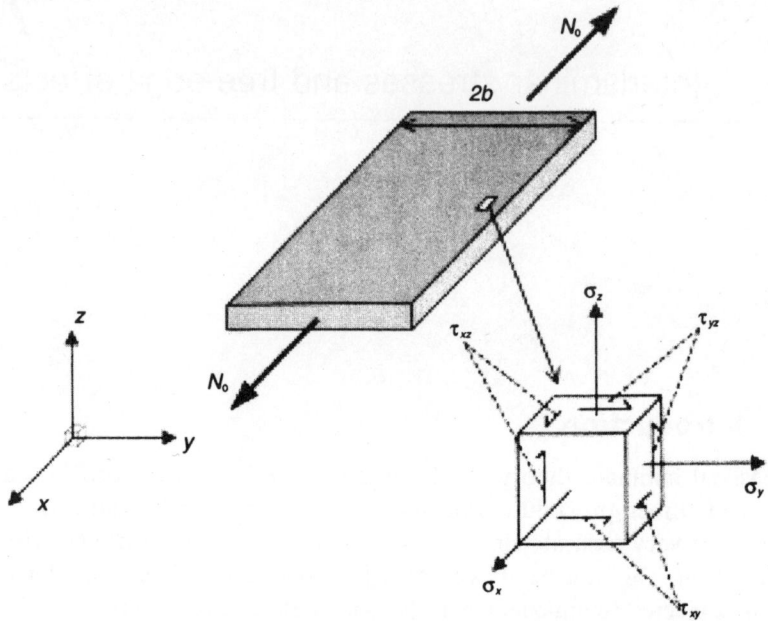

7.1 Stress components developed in a laminate loaded in tension in the x-direction.

from the laminate boundaries. Therefore, the laminate analysis already presented is quite adequate in the regions away from the boundary. However, the state of stress near a free boundary is not a plane stress state but a 3D stress state.

The general principles can be demonstrated by reference to the cross-ply laminate with $(0/90/0°)_s$ lay-up illustrated in Fig. 7.2. The laminate is loaded in tension in the longitudinal direction. The lateral, y-direction, contraction of the 0° layers is opposed by the high lateral stiffness of the 90° layer. As predicted by CLT, this results in tensile σ_y in the 0° layers, counterbalanced by compressive σ_y in the 90° layer; Fig. 7.2(a). The forces acting on a small element of material are shown in Fig. 7.2(b). To satisfy lateral equilibrium at the free edge of the laminate, the resultant force due to these σ_y stresses (forces 1 and 2 in Fig. 7.2b), must be opposed by the resultant of the interlaminar shear stress, τ_{yz} (force 3 in Fig. 7.2b). These forces form a couple which is opposed by the resultant of the direct stress, σ_z (forces 4 and 5 in Fig. 7.2b) near the free edge.[1]

In many cases the transverse stresses, especially the transverse shear stress, may be quite large near the edge and may influence the failure of

| 0° layer |
| 90° layer |
| 0° layer | **(a)** |

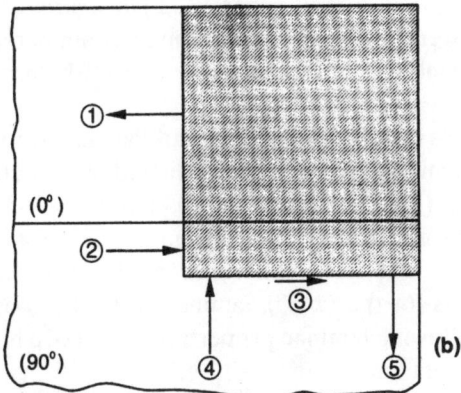

Forces on the shaded element:

① Tensile force due to σ_y in 0° layer
② Compressive force due to σ_y in 90° layer
③ Shear force due to τ_{yz}
④ Compressive force due to σ_z in 0° near laminate edge
⑤ Tensile force due to σ_z in 0° at laminate edge

7.2 Details of the stresses on an element near the edge of a cross-ply laminate.[1]

laminates. A large interlaminar shear stress at the interface may produce matrix cracks at the free edge. These cracks then propagate into the laminate and initiate delaminations, leading to stiffness loss and premature failure of the laminate. This initial damage at the laminate edge is quite important for fatigue loading in which the ultimate failure may initiate at the edges. In this chapter the methods of evaluation of the interlaminar stresses and their influence on the ultimate fracture of the laminates are discussed.

7.2   Development of classical analysis

The interlaminar stress distributions at the free edge of a finite width angle-ply laminate have been examined by many investigators in recent years. Although some of the related studies, which later led to important developments on the subject, were reported by Pagano and Halpin,[2] Pagano,[3,4] Whitney and Leissa[5] and Whitney,[6] the first direct approach to the problem was made by Puppo and Evensen,[7] who derived an approximate formulation in which each of the anisotropic laminae of the laminate were represented by a model that contained an anisotropic plane stress layer and an isotropic shear layer. Each anisotropic layer was assumed to carry only in-plane loads and to exist in a state of generalised plane stress, whereas the isotropic shear layers were assumed to carry the interlaminar shear stress. The interlaminar normal stress was assumed to vanish throughout the laminate.

A second solution was developed by Pipes and Pagano,[8] who considered a four-ply symmetric laminate with the plies oriented only at $\pm\theta$ to the longitudinal laminate axis (Fig. 7.1). The exact equations of elasticity were derived from a uniform axial extension by assuming the stress components to be independent of x. The finite difference method was employed to obtain numerical results for the $(\pm 45°)_s$ laminate with the geometric relation $b = 8h_0$, and the following laminae properties, typical of a high modulus carbon fibre/epoxy system:

$$E_{11} = 138\,\text{GPa} \quad \upsilon_{12} = 0.21$$
$$E_{22} = 14.5\,\text{Gpa} \quad G_{12} = 5.8\,\text{GPa}$$

The stress distributions across the half-width (b) of the specimen are shown in Fig. 7.3. The stresses in the centre of the cross-section are the same as those predicted by CLT. However, as the free edge is approached, σ_x decreases, τ_{xy} goes to zero, and, most significantly, τ_{xz} increases from zero to infinity (a singularity exists at $y = \pm b$). The stresses σ_z and τ_{yz} also increase near the free edge, but their magnitudes are quite small. By use of other laminate geometries, it was established that the width of the region in which the stresses differ from those predicted by lamination theory is approximately equal to the thickness of the laminate (h). Thus the interlaminar stresses, or the deviations from lamination theory, can be regarded as an edge effect only. They are sometimes also referred to as *boundary layer stresses*. It must also be expected that the edge effects will be observed at cut-outs or holes in laminates which provide internal free edges.

The theoretical results of Pipes and Pagano[8] were confirmed experimentally by Pipes and Daniels.[9] The surface displacements of the symmetric angle-ply laminate subjected to axial tension were examined by employing the Moiré technique. The experimental study also confirmed that the inter-

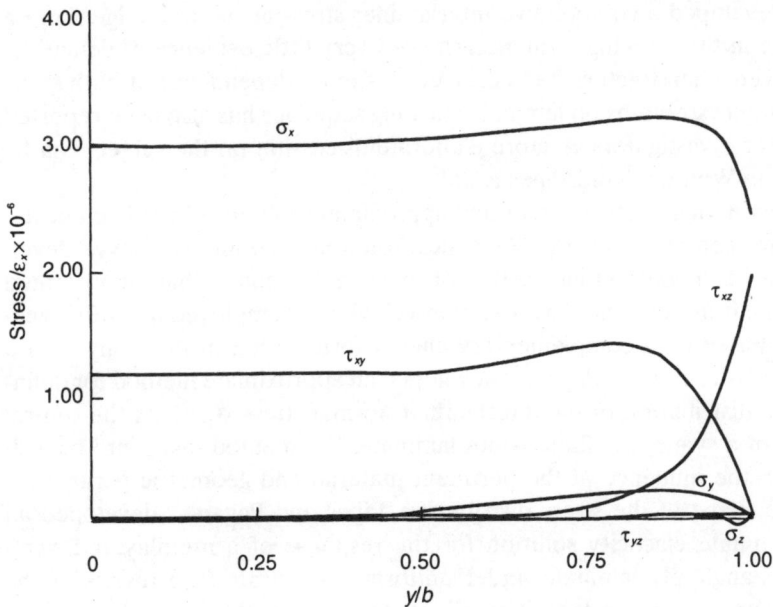

7.3 Stress variations across the width of a (±45°)$_s$ laminate; after Pipes and Pagano.[8]

laminar stresses can be regarded as an edge effect only since their effect is confined to a region whose width is approximately equal to the laminate thickness.

Pagano and Pipes demonstrated in a later study[10] that the interlaminar stresses can be significantly influenced by the laminate stacking sequence, and thus the stacking sequence may be important to a designer. Their work was motivated by observations of Foye and Baker[11] on the tensile fatigue strength of combined angle-ply, (±15/±45°)$_s$ and (±45/±15°)$_s$, boron fibre/epoxy laminates. Foye and Baker reported that fatigue strength of laminates with the former stacking sequence is about 175 MPa lower than that for the latter stacking sequence. Pagano and Pipes showed that the interlaminar normal stress, σ_z, changes from tension to compression by changing the stacking sequence and thus accounts for the difference in strengths. The explanation seems quite reasonable in view of the fact that Foye and Baker observed delamination, and stated that progressive delamination was the failure mode in fatigue. Whitney and Browning[12] and Whitney[13] have also suggested that the interlaminar normal stress σ_z strongly influences the delamination process during failure. During fatigue tests on carbon fibre/epoxy laminates, Whitney observed that a specimen that developed a tensile value of interlaminar stress showed delamination well before final fracture of the specimen. In contrast, another specimen that developed a compressive interlaminar stress at the free edge, from a change in the stacking sequence, showed very little evidence of delamination even when fracture had occurred. A similar dependence of both static and fatigue strengths on laminate stacking sequence has also been reported by other investigators. A more elaborate discussion on the subject can be found in Whitney[13] and Pipes et al.[14]

Many numerical techniques and approximate solutions have been developed for comparison with the classical solutions. Isakson and Levy[15] developed a FE model, similar to that of Pipes and Pagano,[8] that incorporated non-linear interlaminar shear response. Rybicki[16] employed a 3D FE technique based on a complementary energy formulation in the analysis of a finite width laminate. Pagano[17] developed an approximate method for defining the distribution of the interlaminar normal stress, σ_z, along the central plane of a symmetric finite width laminate. The method takes into consideration the influence of the pertinent material and geometric parameters on the shape of the stress distribution. Pipes and Pagano[18] developed an approximate elasticity solution for the response of a multilayered, symmetric, angle-ply laminate under uniform axial strain. The results of the approximate solution exhibit excellent agreement with their earlier numerical results from the exact elasticity equations.[8] Tang[19] obtained an analytical solution for bending of a rectangular composite plate subjected to uniform transverse loading.

Whitney[13] developed an approximate solution based on the numerical results of Pipes and Pagano.[8] Whitney's approximate solution is quite simple to apply and compares reasonably well with the exact elasticity solution.[8] Whitney's solution is discussed in the following paragraphs.

Whitney considered a tensile specimen of length a, thickness h and width b where $b > 2h$. A standard x, y, z coordinate system is located at the mid-plane of the free edge. If the origin is in the gauge section (i.e. away from the ends where the load is applied), the stresses can be assumed to be independent of x. Then, the equilibrium equations take the form

$$\tau_{xy,y} + \tau_{xz,z} = 0$$
$$\sigma_{y,y} + \tau_{yz,z} = 0 \qquad\qquad [7.3]$$
$$\tau_{yz,y} + \sigma_{z,z} = 0$$

where a comma denotes partial differentiation. Based on the numerical results of Pipes and Pagano,[8] Whitney suggested the following form of σ_y and τ_{xy} in the free edge interval $0 \leq y \leq h$:

$$\sigma_y = \frac{\sigma_y(z)}{c}\left[1 - e^{-kn\bar{y}}\frac{k}{n}(\sin n\pi\bar{y} + \cos n\pi\bar{y})\right]$$
$$\tau_{xy} = \frac{\tau_x(z)}{c}(1 - e^{-kn\bar{y}}\cos n\pi\bar{y}) \qquad\qquad [7.4]$$

where

$$c = \left[1 - (-1)^n e^{-k\pi}\right]$$

$$\bar{y} = \frac{y}{h}, \; k > 0$$

n is positive integer, and $\sigma_y(z)$, and $\tau_{xy}(z)$ are determined from lamination theory (equation [7.2]). Substituting equation [7.4] into equation [7.3] and then integrating with respect to z yields the remaining stresses:

$$\tau_{yz} = \frac{-\pi\tau_{yz}(z)(n^2 + k^2)}{nc} e^{-k\pi\bar{y}} \sin n\pi\bar{y}$$

$$\sigma_z = \frac{\pi^2\sigma_z(z)(n^2 + k^2)e^{-k\bar{y}}}{nc}(n\cos n\pi\bar{y} - k\sin n\pi\bar{y}) \qquad [7.5]$$

$$\tau_{xz} = \frac{-\pi\tau_{xz}(z)e^{-ky}}{nc}(n\sin n\pi\bar{y} + k\cos n\pi\bar{y})$$

where

$$[\tau_{yz}(z), \sigma_z(z), \tau_{xz}(z)] = -\int_{-h/2}^{z}[\sigma_y(z), \tau_{yz}(z), \tau_{xy}(z)]_{,y}\mathrm{d}z$$

Thus, equations [7.4] and [7.5] exactly satisfy the equilibrium equations as well as the free-edge boundary conditions. In addition, lamination theory is exactly recovered at $y/h = 1.0$. Compatibility, however, is violated. The solution of these approximate functions compares reasonably well with the exact solution obtained by Pipes and Pagano[8] (shown in Fig. 7.3). The approximate results were obtained with $n = 1$ and $k = 2$. Whitney suggested that since the character of the solution is reasonably approximated with these values of n and k, in the absence of other information they may be used for general application.

7.3   Finite element analysis of straight edges

To illustrate the approach using FE, we consider a four-ply symmetric laminate with $(0/90°)_s$ stacking sequence subjected to an in-plane uniaxial tensile load, as shown in Fig. 7.4(a). The x–y plane of the Cartesian coordinate system lies in the mid-plane of the laminate; the x-dimension is assumed to be much larger than the other dimensions of the plate. The plate width is four times the ply thickness. The 0/90° lay-up is selected for this analysis because of the large mismatch of mechanical properties between layers. The FE77 FE package, developed at Imperial College, Department of Aeronautics,[20] is used, and the analysis is based on isoparametric eight-node elements. Owing to the symmetry of the loading and lay-up, only one-eighth of the laminate is modelled. Because high interlaminar stress

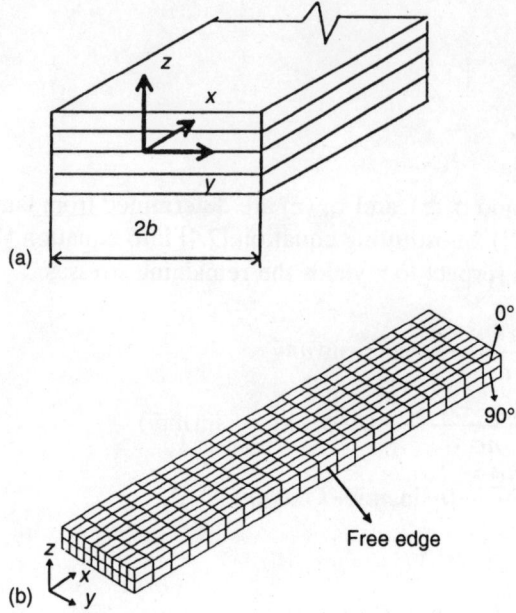

7.4 (a) Cross-ply (0/90°)$_s$ laminate; (b) associated FE mesh.

concentration is expected near the free edge, and at the 0/90° interface, a refined mesh is required in this area; Fig. 7.4(b). Each ply is treated as a homogeneous, elastic and orthotropic material with the following elastic properties: $E_{11} = 146.9\,\text{GPa}$, $E_{22} = E_{33} = 10.89\,\text{GPa}$, $G_{12} = G_{13} = 10.89\,\text{GPa}$, $G_{23} = 6.4\,\text{GPa}$, $\upsilon_{12} = \upsilon_{13} = 0.38$ and $\upsilon_{23} = 0.776$. A uniform tensile strain, $\varepsilon_0 = 10^{-6}$, is applied in the x-direction.

The distributions of the through-thickness direct stress, σ_z, on the laminate midplane and on the 0/90° interface along the y-axis are plotted in Fig. 7.5(a) and (b), respectively. A good agreement between the current model and previous work[8,21] is achieved near the laminate free edge ($0 \leq y/b \leq 0.9$). For a small distance, $0.7 \leq y/b \leq 0.8$, there is a larger difference between the models, probably due to a different type of element and more refined mesh used in Wang and Crossman.[21] Overall, the FE solution provides reasonable results for predicting the through-thickness stresses.

7.4   Finite element analysis of curved edges

7.4.1 Introduction

Cut-outs and holes are widely used in carbon fibre-reinforced laminates for aircraft structures. The existence of the associated stress concentration will

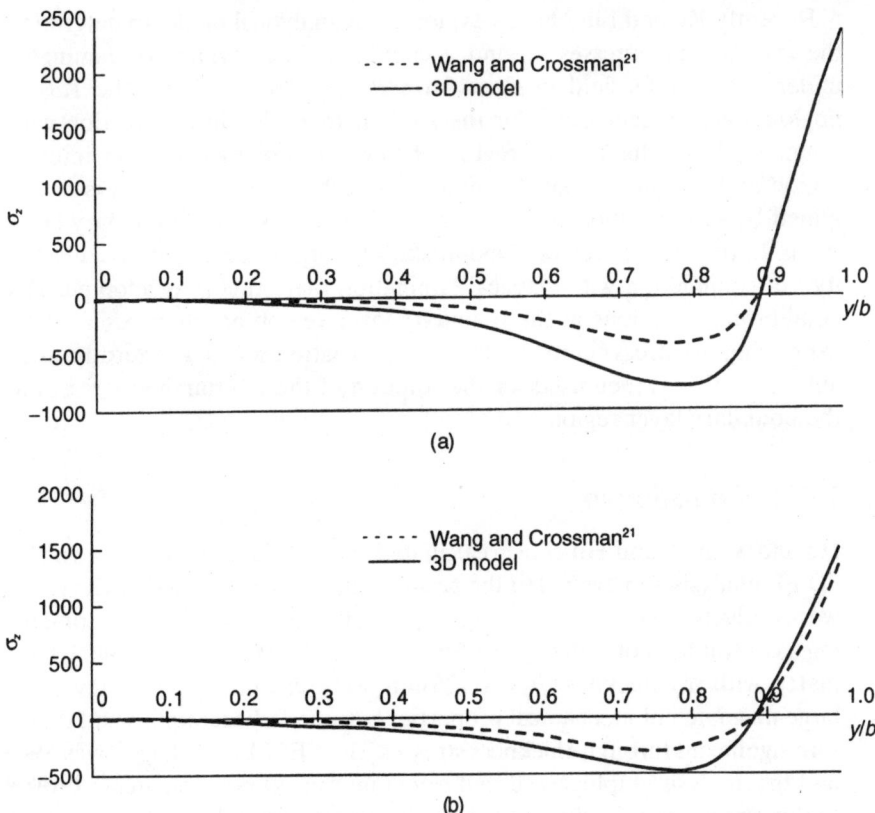

7.5 Variation of through-thickness tensile stress across the width of the cross-ply laminate: (a) at laminate mid-plane, (b) at 0/90° interface.

initiate matrix cracking and delamination at the hole edge, especially under fatigue loading, which will lead to the loss of strength/stiffness and to potential environmental attack.

It has been seen above that the straight free edge effect has been under consideration for more than 20 years. Because of the simplicity of the straight free edge geometry as compared with a curved free edge, fewer studies have been done on the latter. The straight free edge can sometimes be analysed as a problem with only 2D variations in stresses and displacements. Unfortunately, the stress state near a circular hole in a laminate is a complex 3D problem. The only exact solution to the problem is a 2D approximation.[22] Approximate analytical methods[23,24] and numerical approaches such as finite difference and FE methods[25-27] have been used with some success to analyse the interlaminar stress distributions around curved free edges.

Recently, Ko and Lin[28] have developed an analytical model to determine the interlaminar stresses around a circular hole in symmetric laminates under a set of far-field in-plane stresses. This method uses the Kassapoglou–Lagace technique[29] for the straight free edge in conjunction with boundary layer theory. The region of interest is divided into an interior region and a boundary layer region, and each stress component is determined by superposition of the interior stress field and the boundary layer stress. In the interior region, Lekhnitskii's theory of 2D anisotropic elasticity[22] in conjunction with classical lamination plate theory is adopted. The equilibrium equations in the boundary layer region are approximated by expressing the stress components as a perturbation series. The zeroth-order approximation is then used for the solution of the interlaminar stresses in the boundary layer region.

7.4.2 An example

Hu and Soutis[30] and Hu *et al.*[31] calculated the interlaminar stresses using a 3D FE analysis and evaluated the accuracy of the Ko–Lin model. The study was conducted on a plate of length $L = 60$ mm, width $w = 30$ mm, containing a central hole of radius $R = 2.5$ mm. Cross-ply $(90/0°)_s$ and $(0/90°)_s$ laminates, with ply thickness $h_0 = 0.125$ mm, were examined since they show large mismatch of mechanical properties between adjacent plies and therefore significant through-thickness stresses. The FE77 FE package[20] was used and the analysis employed curved isoparametric 20-node elements. Owing to the symmetry of loading, hole location and lay-up, only one-eighth of the laminate was modelled. Because high interlaminar stress concentration is expected near the hole edge, and at the 0/90° interface, a very high mesh refinement is required in this area. The smallest element dimension in the radial and thickness directions was 0.00025 mm. Figure 7.6 shows the FE mesh used. This FE model consisted of 4000 elements and solved a system of 56000 linear equations. Each ply was treated as a homogeneous, elastic orthotropic material with the same elastic properties as those in the literature of Raju and Crews.[26]

7.4.2.1 Interlaminar normal stresses

Figure 7.7 shows the interlaminar normal stress distributions at the 90/0° ply interface around the hole ($z = h_0 = 0.125$ mm) of the $(90/0°)_s$ laminate. Since a mathematical interlaminar stress singularity exists at the free edge between the 90° and 0° plies, the computed stresses are presented near, but not at, the hole boundary. The stresses closest to the hole are at $r/R = 1.0001$; that is, at a distance of $0.002h_0$ from the hole boundary. As the distance from the edge, $(r–R)$, increases, the interlaminar stress σ_z decreases rapidly.

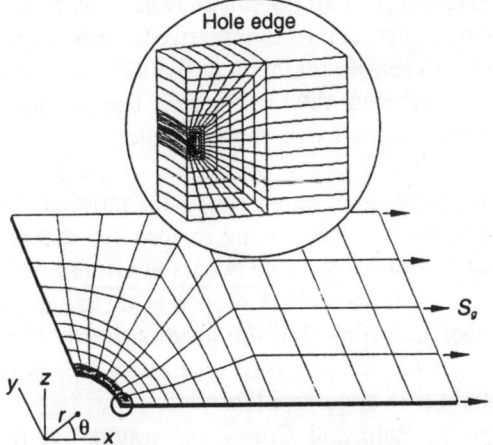

7.6 Three-dimensional finite element mesh for one-eighth the notched (90/0°)$_s$ laminate.[31] (s_g = remote applied stress.)

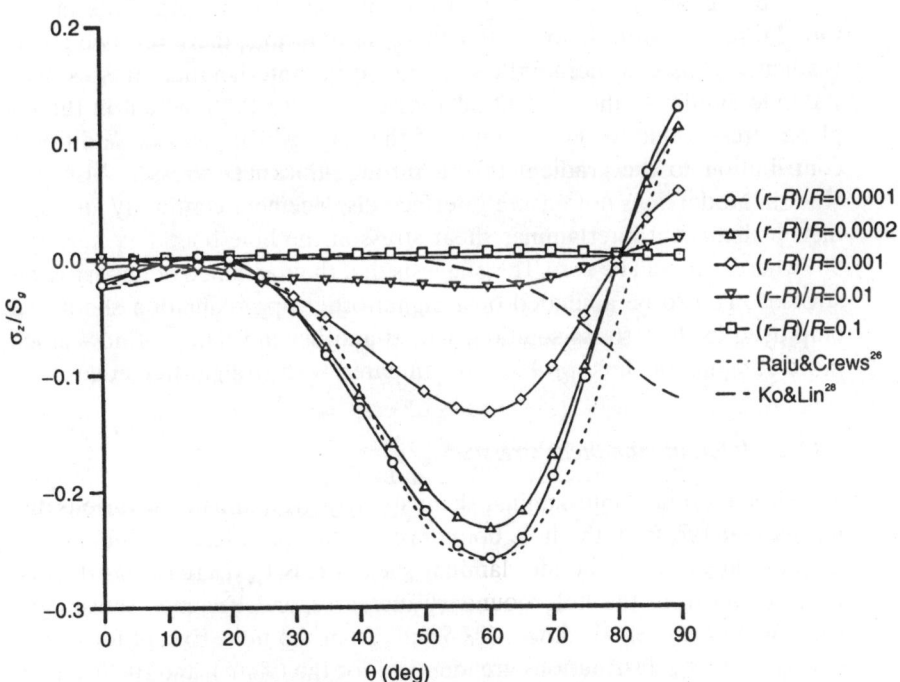

7.7 Normalised interlaminar direct stress distributions around the hole at the 90/0° interface of the (90/0°)$_s$ laminate.[31]

When $(r-R) = 0.1R$, i.e. 0.25 mm, or two ply-thicknesses away from the hole boundary, σ_z becomes almost zero, and is compressive for most of the region around the hole, with a small tensile region near $\theta = 90°$. The largest compressive σ_z occurs at about 60° from the loading axis. The predicted values at the central (0/0°) interface are twice those developed at the (90/0°) interface.

For the (0/90°)$_s$ lay-up, σ_z is again compressive for most of the region around the hole boundary, but is tensile in the regions $10° \leq \theta \leq 28°$ and $80° \leq \theta \leq 90°$. Its largest compressive value occurs at around $\theta = 60°$. The stress distributions in Fig. 7.7 have similar shapes at different $(r-R)$ distances from the hole edge, indicating that the location in the hoop direction is an important parameter for a curved free edge. This is the main difference from the straight free edge problem.

For comparison purposes, Raju and Crews' FE results[26] at $(r-R)/R = 0.0001$ are plotted in Fig. 7.7. The difference between the two FE solutions is negligible. However, the stress predictions of the Ko and Lin[28] analytical model are in poor agreement, both qualitatively and quantitatively. This is probably due to the zeroth-order approximation that those authors used when they solved the equilibrium equations in the boundary layer region. Although their solution satisfied the equilibrium conditions at the boundary, the circumferential stress and the laminate's stress gradients in the radial direction were ignored. For the case of a hole, there are two stress gradients to be considered: the gradient in the interlaminar stresses near the hole, similar to the straight edge case; and also the gradient in the in-plane stresses due to the presence of the hole, which gives an additional contribution to the gradient in the through-thickness stresses. Also, the Ko–Lin model does not ensure interface displacement continuity and predicts a significant interlaminar shear stress at the hole boundary near $\theta = 90°$ where it should be zero. This suggests that their assumed boundary layer stresses need to be amended or a higher-order approximation should be employed in their stress solution. This is another indication of how much more complex the hole problem is compared with straight free edges.

7.4.2.2 Interlaminar shear stresses

The circumferential interlaminar shear stress distributions $\tau_{z\theta}$ at various distances, $(r-R)/R$, from the hole boundary are shown in Fig. 7.8. Similarly to the normal stress, σ_z, the interlaminar shear stress $\tau_{z\theta}$ decreases as the distance $(r-R)$ from the hole boundary increases, and becomes vanishingly small within two ply thickness (0.25 mm) from the hole. Except for different signs, the $\tau_{z\theta}$ distributions are identical for the (90/0°)$_s$ and (0/90°)$_s$ laminates. Both of these distributions have their maximum values at about 75° from the loading axis. The largest value in Fig. 7.8 is approximately $1.65 S_g$

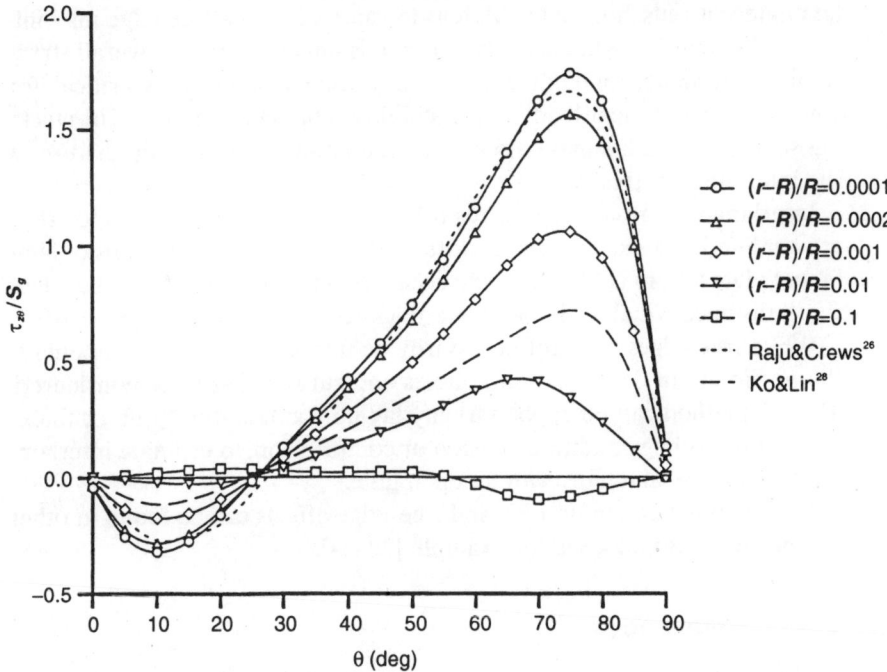

7.8 Normalised interlaminar shear stress distributions around the hole at the 90/0° interface of the (90/0°)$_s$ laminate.[31]

(S_g being the remote applied stress), which is about ten times larger than the largest σ_z value (Fig. 7.7) computed for the same distance from the hole. This comparison indicates that the $\tau_{z\theta}$ stress singularity is stronger than that for the σ_z interlaminar stress. It also suggests that the interlaminar shear stresses are mainly responsible for delamination initiation in such laminates. The shear stress component τ_{zr} is very small compared with $\tau_{z\theta}$ and can be neglected.

Raju and Crews' FE results[26] for $\tau_{z\theta}$ are in very good agreement with the present FE model, while those predicted by the Ko–Lin analytical model[28] are lower by almost a factor of 3. The zeroth-order approximation of the Ko–Lin analysis, although giving a good qualitative agreement with the numerical models, produces considerably lower interlaminar shear stresses and therefore will predict higher delamination onset stresses.

7.5   Summary

Interlaminar normal and shear stresses can be very high (perhaps even singular) at the free edge of a laminate (as at the edges on the sides of a laminate, cut-outs, holes, etc.). A tensile value of σ_z at a free edge may initiate delamination. Whereas delamination is important to the overall structural performance of laminates under tensile loading, it is critical for compression and shear loads where stability is of major concern. The stacking sequence in a laminate affects the magnitude as well as the nature of the interlaminar stresses.

Interlaminar stresses can be regarded as an edge effect only since they are confined to a narrow region close to the edges. The predictions of laminated plate theory (LPT) are quite accurate in regions away from the edges (i.e. a distance equal to the laminate thickness).

While free edges may not necessarily be important for real components the possibility of delamination at holes and cut-outs has to be considered. The FE method can be applied to any lay-up configuration (thin or thick) under any loading condition, tension or compression, to estimate interlaminar stresses in laminates with an open hole.

Further information on LPT and free-edge effects can be found in other composites textbooks, see for example [32–34].

7.6   References

1 Curtis P T, 'The effect of edge stresses on the failure of (0/45/90°) CFRP laminates', *TR 80054*, Farnborough, UK, Royal Aerospace Establishment, April, 1980.

2 Pagano N J & Halpin J C, 'Influence of end constraints in the testing of anisotropic bodies', *J Compos Mater*, 1968 **2**(4) 18–31.

3 Pagano N J, 'Exact solutions for composite laminates in cylindrical bending', *J Compos Mater*, 1969 **3**(3) 398–411.

4 Pagano N J, 'Extract solutions for rectangular bidirectional composites and sandwich plates', *J Compos Mater*, 1970 **4**(1) 20–34.

5 Whitney J M & Leissa A W, 'Analysis of heterogeneous anisotropic plates, *J Appl Mech*, 1969 **28** 261–266.

6 Whitney J M, 'The effects of transverse shear deformation on the bending of laminate plates', *J Compos Mater*, 1969 **3**(4) 534–547.

7 Puppo A H & Evensen H A, 'Interlaminar shear in laminated composites under generalised plane stress', *J Compos Mater*, 1970 **4**(2) 204–220.

8 Pipes R B & Pagano N J, 'Interlaminar stresses in composite laminates under uniform axial extension', *J Compos Mater*, 1970 **4**(4) 538–548.

9 Pipes N J & Daniel I M, 'Moiré analysis of the interlaminar shear edge effect in laminated composites', *J Compos Mater*, 1971 **5**(2) 255–259.

10 Pagano N J & Pipes R B, 'The influence of stacking sequence of laminate strength', *J Compos Mater*, 1971 **5**(1) 50–57.

11 Foye R L & Baker D J, 'Design of orthotropic laminates', presented at the *11th Annual AIAA Structures, Structural Dynamics, and Materials Conference*, Denver, CO., April 1970.

12 Whitney J M & Browning E E, 'Free-edge delamination of tensile coupons', *J Compos Mater*, 1972 **6**(2) 300–303.

13 Whitney J M, 'Free-edge effects in the characterisation of composite materials', in *Analysis of the Test Methods for High Modulus Fibers and Composites*, ASTM STP 521, Philadelphia, PA, American Society for Testing and Materials, 1973.

14 Pipes R B, Kaminski B E & Pagano N J, 'Influence of the free-edge upon the strength of angle-ply laminates', in *Analysis of the Test Methods for High Modulus Fibres and Composites*, ASTM STP 521, Philadelphia, PA, American Society for Testing and Materials, 1973.

15 Isakson G & Levy A, 'Finite element analysis of interlaminar shear in fibrous composites', *J Compos Mater*, 1971 **5**(2) 273–276.

16 Rybicki E F, 'Approximate three-dimensional solutions for symmetric laminates under inplane loading', *J Compos Mater*, 1971 **5**(3) 354–360.

17 Pagano N J, 'On the calculation of interlaminar normal stress in composite laminate', *J Compos Mater*, 1974 **8**(1) 65–81.

18 Pipes R B & Pagano N J, 'Interlaminar stresses in composite laminates – An approximate elasticity solution', *J Appl Mech*, 1974 **41**, Series E(3) 668–672.

19 Tang S, 'Interlaminar stresses of uniformly loaded rectangular composite plates', *J Compos Mater*, 1976 **10**(1) 69–78.

20 Hitchings D, *Finite Element Package FE77: User's Manual*, London, Imperial College, 1996.

21 Wang A S D & Crossman F W, 'Some new results on edge effects in symmetric composite laminates', *J Compos Mater*, 1977 **11**(1) 92–106.

22 Lekhnitsiskii S G, *Theory of Elasticity of an Anisotropic Body*, Moscow, Mir Publishers, 1981.

23 Tang S, 'Interlaminar stresses of uniformly loaded rectangular composite plates', *J Compos Mater*, 1976 **10**(1) 69–78.

24 Tang S, 'A variational approach to edge stresses of circular cut-outs in composites'. *AIAA/ASME/ASCE/AHS/SDM Conference*, St Louis, MO, 326–332, 1979.

25 Rybicki E F & Schmuesser D W, 'Effect of stacking sequence and lay-up angle on free edge stresses around a hole in a laminated plate under tension', *J Compos Mater*, 1978 **12**(7) 300–313.

26 Raju I S & Crews J H, 'Interlaminar stress singularities at a straight free edge in composite laminates', *J Computers Structures*, 1981 **14**(1–2) 21–28.

27 Kim J Y & Hong C S, 'Three-dimensional finite element analysis of interlaminar stresses in thick composite laminates', *J Computers Structures*, 1991 **40**(6) 1395–1404.

28 Ko C C & Lin C C, 'Method for calculating the interlaminar stresses in symmetric laminates containing a circular hole', *AIAA J*, 1992 **30**(1) 197–204.

29 Kassapoglou C & Lagace P A, 'An efficient method for the calculation of interlaminar stresses in composite materials', *ASME J Appl Mech*, 1986 **53**(4) 744–750.

30 Hu F Z & Soutis C, 'Evaluation of the Ko–Lin model for interlaminar stresses in composite laminates with an open hole', *Adv Compos Letters*, 1996 **5**(5) 143–147.

31 Hu F Z, Soutis C & Edge E C, 'Interlaminar stresses in composite laminates with a circular hole', *Compos Struct*, 1997 **37**(2) 223–232.

32 Jones R M, *Mechanics of Composite Materials*, USA, Taylor & Francis, 1975.
33 Agarwal B D & Broutman L J, *Analysis and Performance of Fiber Composites*, USA. J Wiley & Sons, Inc, 1990.
34 Halpin J C, *Primer on Composite Materials Analysis*, USA, Technomic Publishing Company Inc, 1992.

8
Fracture and fracture mechanics

8.1 Introduction

Typical engineering fibre-reinforced plastics consist of brittle fibres, such as glass or carbon, in a weak brittle polymer matrix, such as epoxy or polyester resin. However, an important characteristic of these composites is that they are reasonably tough, largely as a result of their heterogeneous nature, the manner of their construction, and the widespread modes of fracture. During deformation, microstructural damage is extensive throughout the composite, but much damage can be sustained before load-bearing ability is impaired. Beyond some critical level of damage, failure may occur by the propagation of a crack which usually has a much more complex character than cracks in metals. Crack growth is inhibited by the presence of interfaces between fibres and matrix and between separate laminae in a multiply laminate.

Compared with fracture in metals, research into the fracture behaviour of composites is quite limited. Much of the necessary theoretical framework is not yet fully developed and there is no simple recipe for predicting the toughness of all composites. We are not yet able to design with certainty the make-up of any composite so as to produce the optimum combination of strength and toughness. However, it will be apparent from what follows that the lay-up geometry of a composite strongly affects crack propagation, with the result that some laminates appear highly notch-sensitive whereas others are totally insensitive to the presence of stress concentrators. The selection of fibres and resins, the manner in which they are combined in the lay-up, and the quality of the manufactured composite must all be carefully controlled if optimum toughness is to be achieved. Furthermore, requirements for highest tensile and shear strengths of laminates are often incompatible with requirements for highest toughness. Final selection of a composite for a given application may therefore be a matter of compromise.[1]

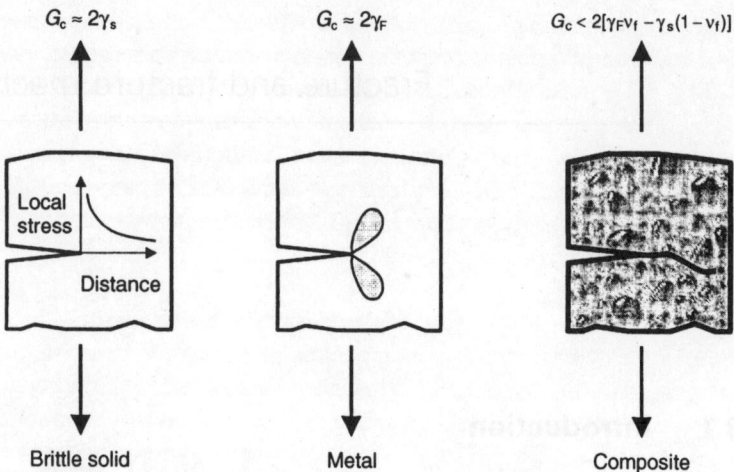

8.1 Schematic illustration of cracking mechanisms in different kinds of solids.[1]

8.2   Crack tip behaviour in homogeneous solids

The principal differences between cracking mechanisms in some familiar solids are shown in Fig. 8.1.[1] In an ideal brittle solid, failure is elastic by the simple breaking of bonds across the crack plane, which remains relatively flat. The work of fracturing the solid in terms of the familiar Griffith model[2] is approximately the work of creating two new surfaces:

$$G \cong 2\gamma_s = \frac{\pi\sigma^2 a}{E} \qquad [8.1]$$

where γ_s = specific surface energy, σ = applied stress, a = one half crack length and E = modulus of elasticity.

Griffith rewrote equation [8.1] in the form:

$$\sigma = \left(\frac{2E\gamma_s}{\pi a}\right)^{\frac{1}{2}} \qquad [8.2a]$$

for the case of plane stress (biaxial stress conditions), and:

$$\sigma = \left[\frac{2E\gamma_s}{\pi a(1-\upsilon^2)}\right]^{\frac{1}{2}} \qquad [8.2b]$$

for the case of plane strain (triaxial stress conditions associated with the suppression of strains in one direction). The Poisson ratio, υ, for most materials is between 0.25 and 0.33.

It is important to recognise that the Griffith relation was derived for an ideal elastic material containing a very sharp crack. Although equations [8.2a] and [8.2b] do not explicitly involve the crack tip radius, ρ, that radius is assumed to be very sharp. As such, the Griffith relation, as written, should be considered necessary but not sufficient for failure. The crack tip radius would also have to be atomically sharp to raise the local stress above the cohesive strength.

In materials such as metals and plastics, even relatively brittle ones, energy is dissipated in non-elastic deformation mechanisms in the region of the crack tip. This energy is lost in moving dislocations in a metal and in viscoelastic flow or craze formation in a polymer. The zone around the crack tip in which these deformation processes occur is called the 'process zone'. In such real cases the fracture energy, G, is found to be several orders of magnitude greater that the material surface energy γ_s. Orowan[3] recognised this fact and suggested that equation [8.1] be modified to include the energy of plastic deformation γ_p in the fracture surface, so that:

$$G = 2(\gamma_s + \gamma_p) = 2\gamma_F \qquad [8.3]$$

where γ_F is known as the 'work of fracture'.

Irwin[4] also considered the application of Griffith's relation, equation [8.2a], to the case of materials capable of plastic deformation. Instead of developing an explicit relation in terms of the energy sink terms, γ_s or $(\gamma_s + \gamma_p)$, Irwin chose to use the energy source term (i.e. the elastic energy per unit increment of crack length, $\partial U/\partial a$). Denoting $\partial U/\partial a$ as G, Irwin showed that a microcrack, or growth defect, may propagate at a nominal applied failure stress given by:

$$\sigma_f \cong \left(\frac{EG_{Ic}}{\pi a}\right)^{\frac{1}{2}} = \frac{K_{Ic}}{\sqrt{\pi a}} \qquad [8.4]$$

where G_{Ic} and K_{Ic} are the critical elastic energy release rate and fracture toughness, respectively; $K_{Ic}^2 = EG_{Ic}$ for isotropic materials. The K_{Ic} parameter is a material property and can be measured in the laboratory with sharply notched test specimens. Equation [8.4] is one of the most important relations in the literature of linear elastic fracture mechanics (LEFM).

Fracture mechanics has been used successfully in the development of structural materials in two complementary ways: firstly, by relating K_{Ic} to microstructure; secondly, by identifying the origins of cracking. The strength of a component can then be maximised by optimising the microstructure through careful control of manufacturing conditions to reduce flaw size and increase K_{Ic}. This requires an understanding of the dependence of K_{Ic} (or G_{Ic}) on microstructure.

In general, materials that show some ductility can fail in one of two competing ways, depending on the site of any crack or notch they may contain. A material containing a short notch fails by general damage, in which ductile fracture occurs at a net section stress which is independent of notch length. Alternatively, the localised stresses at the tip of a long notch or crack may be large enough to cause crack propagation, while deformation in the bulk of the material remains elastic. In this case fracture occurs at a stress that is proportional to $1/\sqrt{a}$ (equation [8.4]). We call this the regime of single crack propagation.

In composite materials, the fibres interfere with crack growth, but their effect depends on how strongly they are bonded to the matrix (resin). For example, if the fibre/matrix bond is strong, the crack may run through both fibres and matrix without deviation, in which case the composite toughness would be low and approximately equal to the sum of the separate constituents' toughness:

$$G \cong 2[(\gamma_F)_f v_f + (\gamma_F)_m (1 - v_f)]$$

[8.5]

where $(\gamma_F)_f$ and $(\gamma_F)_m$ are the fibre and matrix work of fracture, respectively. The fibre volume fraction is v_f.

On the other hand, if the fibre/matrix bond is weak the crack path becomes very complex and many separate damage mechanisms may then contribute to the overall fracture work of the composite. These fracture processes are controlled by the constituent materials of the composite and the manner of their combination. For example, a brittle polymer or epoxy resin with $G_{Ic} \cong 0.1\,\text{kJ/m}^2$ and brittle glass fibres with $G_{Ic} \cong 0.01\,\text{kJ/m}^2$ can be combined together in composites some of which have energies of up to $100\,\text{kJ/m}^2$. For an explanation of such a large effect we must look beyond simple addition.

8.3   Crack extension in composites

Many authors have attempted to apply fracture mechanics to fibre-reinforced composites and have met with mixed success. Conventional fracture mechanics methodology assumes a single dominant crack that grows in a self-similar fashion; i.e. the crack increases in size (either through stable or unstable growth), but its shape and orientation remain the same. Fracture of a fibrous composite, however, is often controlled by numerous microcracks distributed throughout the material, rather than a single macroscopic crack. There are situations where fracture mechanics is appropriate for composites, e.g. for delamination, but it is important to recognise the limitations of theories that were intended for homogeneous materials. Figure 8.2[5] illustrates various failure mechanisms in fibre-reinforced

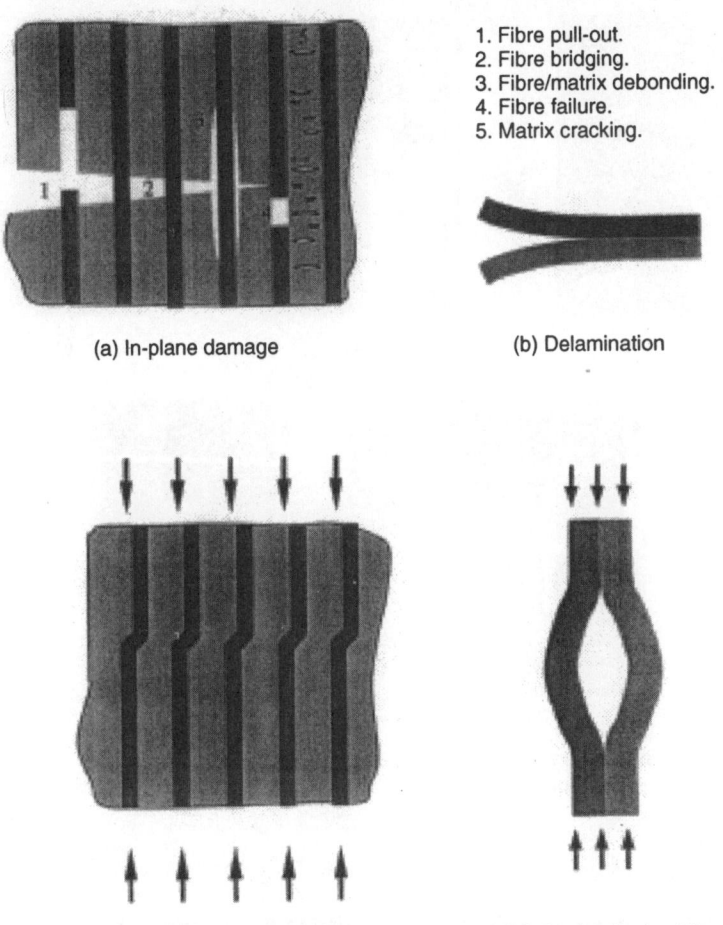

1. Fibre pull-out.
2. Fibre bridging.
3. Fibre/matrix debonding.
4. Fibre failure.
5. Matrix cracking.

(a) In-plane damage

(b) Delamination

(c) Fibre microbuckling

(d) Delamination buckling

8.2 Examples of damage and fracture mechanisms in fibre-reinforced composites.[5]

composites. One advantage of composite materials is that fracture seldom occurs catastrophically without warning, but tends to be progressive, with substantial damage widely dispersed throughout the material. Tensile loading, Fig. 8.2(a), can produce matrix cracking, fibre bridging, fibre rupture, fibre pull-out and fibre/matrix debonding. Ultimate tensile failure of a fibrous composite often involves several of these mechanisms. Out-of-plane stresses can lead to delamination, Fig. 8.2(b), because the fibres do not contribute significantly to strength in this direction. Compressive loading can produce microbuckling of fibres, Fig. 8.2(c); since the polymer matrix is soft compared with the fibres, the fibres are unstable in compres-

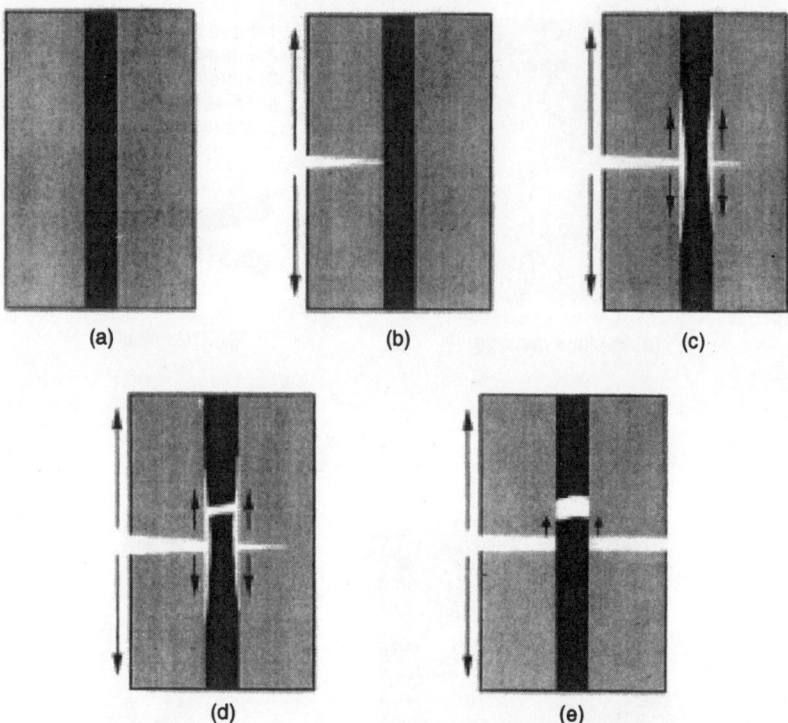

8.3 Schematic representation of stages in crack growth in a fibre composite.[1]

sion. Compressive loading can lead also to macroscopic delamination buckling, Fig. 8.2(d), particularly if the material contains a pre-existing delaminated region. All these failure mechanisms absorb energy and contribute
to the fracture toughness of the composite.

In order to distinguish the separate micromechanisms of toughening due
to true composite action, it is convenient to consider a simple model in
which a crack travelling in the resin approaches an isolated fibre; Fig. 8.3a.[1]
A crack propagating in the matrix phase is effectively stopped by a fibre;
Fig. 8.3(b). As the load on the composite is increased, matrix and fibre at
the crack tip deform differentially and a large local stress builds up in the
fibre. This stress causes local Poisson contraction which, exacerbated by
the tensile stress normal to the interface ahead of the crack tip, initiates
fibre/resin debonding (or decohesion); Fig. 8.3(c). The interfacial shear
stress resulting from fibre/matrix modulus mismatch will then cause extension of the debond along the fibre in both directions away from the crack

plane. This permits further opening of the matrix crack beyond the fibre, and the process is repeated at the next fibre. An upper limit to the energy of debonding is given by:

$$W_{debond} = \frac{N\pi d^2 \sigma_f^2 y}{8E_f} \qquad [8.6]$$

for a composite with N fibres of diameter d, failure stress σ_f and modulus E_f. The mean bonded length is y.

After debonding, the fibre and matrix move relative to each other as crack opening continues and work must be done against frictional resistance during this process. One estimate, assuming that the interfacial frictional stress, τ, acts over a distance equal to the fibre failure extension, suggests that this contribution is equal to:

$$W_{friction} = \frac{N\pi dy^2 \pi \varepsilon_f}{2} \qquad [8.7]$$

where ε_f, the fibre failure strain, contributes substantially to the toughness of fibre/resin composites.

After debonding a continuous fibre is loaded to failure over a distance equal to the debonded length and it may break at any point within this region; Fig. 8.3(d). The broken ends retract and regain their original diameters and they are regripped by the resin. In order to permit further opening of the crack, and ultimately separate the two parts of the sample, these broken ends must be pulled out of the matrix; Fig. 8.3(e). Further frictional work is needed to accomplish this. Crude estimates of the pull-out work give:

$$W_{pull-out} = \frac{N\pi d l_c^2 \tau}{12} \qquad [8.8]$$

where l_c is the critical length, and the distance over which the fibre end is pulled out is given approximately by $l_c/4$. It can be shown that, in aligned short fibre composites, the work of pull-out will be at a maximum when the reinforcing fibres are exactly of the critical length.

It should be noted that composite toughness ought to be increased by raising the fibre volume fraction, v_f, increasing the fibre diameter, d, or using stronger fibres. Contrary to expectation, however, improving the fibre/matrix bond will usually reduce the toughness because it inhibits debonding and therefore reduces pull-out.

The behaviour of many types of composite has been reasonably well explained in terms of summation of the contributions from these mechanisms but it is not yet possible to design a composite to have a given tough-ness. Clearly, ease of cracking in unidirectional laminates will be strongly dependent on fibre orientation. In GRP and CFRP, for example, the energy for fracture parallel to the fibres is two orders of magnitude less than that normal to the fibres. The same is true of wood.

Simple crack interactions rarely occur in practical composites made by bonding together several plies with differing fibre orientations. The varia-tion of axes of anisotropy from lamina to lamina results in coupling shear stresses in the plane of the plate when the composite is loaded and, since the interlaminar planes are always planes of weakness, it follows that inter-laminar shear stresses may easily become so large as to delaminate the com-posite well before fibre fracture can occur. Cracks are also caused to deviate along these planes of weakness. Because of the increased surface area asso-ciated with this form of cracking the composite toughness is substantially increased. Delamination can be considered as a special case, and is dealt with separately in the next chapter.

8.4   Application of fracture mechanics to composites

8.4.1   Notched strength and notch sensitivity

Frequently the design of composite structures includes holes introduced either intentionally as cut-outs and as fastener holes, or unintentionally due to damage events. Several investigators (e.g. Waddoups et al.[6] and Soutis et al.[7]) have examined the behaviour of notched composite laminates and found that open holes reduced the tensile or compressive strength of the laminate by more than 50%. However, the remote failure stress is well above the value one might predict from the elastic stress concentration factor, indicating that the composite material is not ideally brittle and some stress relief occurs around the hole.[7,8] In Fig. 8.4 the remote compressive failure stress, σ_n, normalised by the unnotched failure strength, σ_{un}, of the laminate is plotted as a function of hole radius R normalised by semi-width W of the specimen. The high stresses at the hole boundary initiate localised damage which results in a redistribution of stresses. Such damage will take the form of delamination, matrix cracking, splitting, fibre fracture (fibre microbuckling in compression). It is reasonable to expect that fast fracture of the laminate will occur when the size of the delaminations, or the length of the crack tip split, or size of the damage zone, exceeds some critical value. Various theoretical models have been developed to predict notched tensile fracture.[6,9] Some of the models have been extended to include compression-loaded laminates.[10] More recently the progressive development of fibre microbuckling leading to ultimate compressive failure has been modelled successfully by Soutis and coworkers.[7,8]

L1: $[(\pm 45/0_4)_2]_s$, L2: $[(\pm 45/0_2)_3]_s$, L3: $[(0/90_2/0)_3]_s$, L4: $[\pm 45/0_2/90_2/0_2/90_2/0_2]_s$,
L5: $[(\pm 45/0/90)_3]_s$, L6: $[(\pm 45)_2/0/(\pm 45)_2/0/\pm 45]_s$

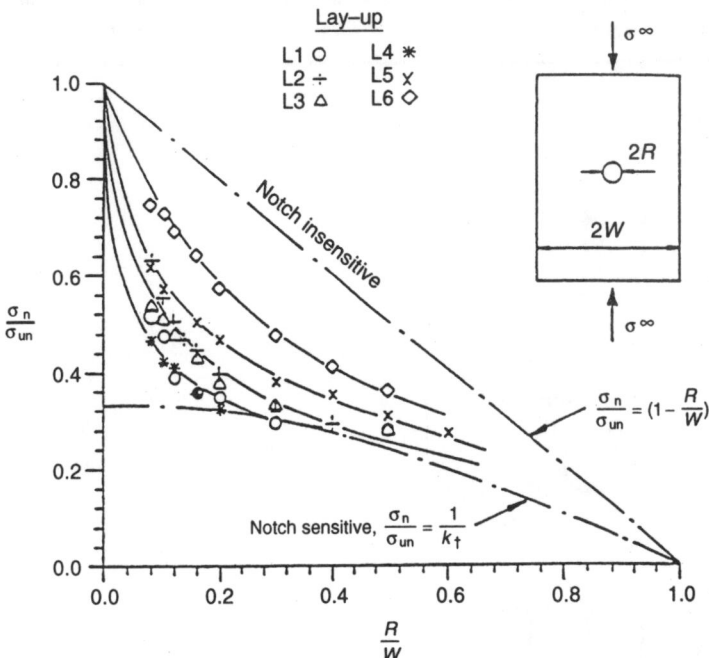

8.4 Effects of hole diameter on the compressive strength of T800/924C
laminates.[8]

8.4.2 Inherent flaw model

Waddoups *et al.*[6] postulated the existence of a 'high-intensity energy region',
adjacent to a circular hole in a composite plate under uniaxial tensile load;
Fig. 8.5. The high intensity energy region is presumed to behave like a crack
of length a_0. The fracture strength of the laminate σ_n, is taken from the
toughness K_{Ic} and a_0, given the geometry and stress intensity factor, K_{Ic}.
Waddoups made use of Bowie's solution[11] for the stress intensity factor,
for the problem of cracks emanating from a circular hole in an *isotropic*
homogeneous infinite plate.

It was shown that the flawed, or notched, strength (σ_n) and unflawed, or
unnotched, strength (σ_{un}) of the composite are related by:

$$\frac{\sigma_{un}}{\sigma_n} = f(a_0/r) \qquad\qquad [8.9]$$

where the correction factor, $f(a_0/r)$, from Bowie is given in Table 8.1.
Waddoups considered a $(0/\pm 45°)_{2s}$ carbon fibre/epoxy laminate containing

Table 8.1 Correction factors $f(a_o/r)$[11]

a/r	0.1	0.2	0.3	0.4	0.5	0.6	0.8	1.0
f(a₀/r)	2.73	2.41	2.15	1.96	1.83	1.71	1.58	1.45

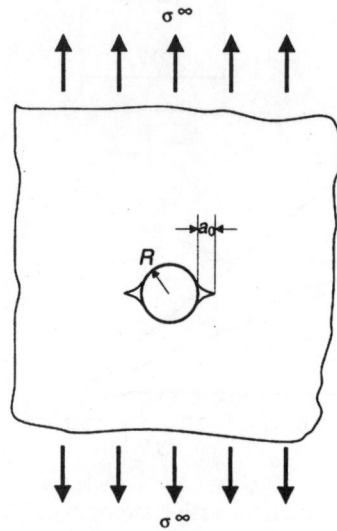

8.5 Damage zone of size a_o, at edge of a hole or cut-out.[6]

a circular hole, the strength data for which are given in Table 8.2. Inserting values of $\sigma_{un} = 524\,\text{MPa}$ and $\sigma_n = 192\,\text{MPa}$ into equation (8.9) gives:

$$f(a_0/r) = \frac{\sigma_{un}}{\sigma_n} = \frac{524}{192} = 2.72$$

From Table 8.1 we see that $a_0/r = 0.1$ and therefore $a_0 = 1.27$. We now calculate K_{Ic} using:

$$K_{Ic} = \sigma_n (\pi a_0)^{\frac{1}{2}} f(a_0/r) \cong 31\,\text{MPa}\sqrt{m}$$

Assuming that K_{Ic} is a laminate property, we can then determine the notched strength for specimens containing holes of diameters 63 mm and 75 mm. Values of σ_n calculated this way are 166 MPa and 164 MPa, respectively, which compare favourably with experimental measurements of 157 MPa and 156 MPa given in Table 8.2.

Table 8.2 Hole data summary[6]

Specimen (965 mm × 125 mm)	Static strength (MPa)
Control (no hole)	524
25 mm diameter hole	192
63 mm diameter hole	157
75 mm diameter hole	156

8.4.3 Stress intensity factors for a cracked hole laminate

For isotropic materials the subject of linear elastic fracture mechanics (LEFM) is highly developed and the plane strain fracture toughness, K_{Ic}, is treated as a material property. Paris and Sih[12] developed expressions for the crack tip stress field for a linear elastic anisotropic material for the case of plane strain and pure shear. The crack tip stress field exhibits the same $r^{-1/2}$ singularity with distance r from the crack tip as for the isotropic case. Also the stress intensity factors (SIF), K_I, K_{II}, K_{III}, have the same meaning as for the isotropic case.

Linear elastic fracture mechanics can be applied to composite laminates provided the damage zone at the crack tip is contained within the K-field, i.e. provided the damage zone is small compared with other dimensions of the specimen.

The stress intensity factor for a finite 2D isotropic plate, subjected to mode I loading and containing a cracked hole, Fig. 8.5, is given by

$$K_I = \sigma^\infty \sqrt{\pi} (r + a_0) f(a_0/r) F(r, a_0, w) \qquad [8.10]$$

where σ^∞ is the applied far field stress and the factor $f(a_0/r)$ reflects the influence of the cracked hole on the SIF. Function F is the finite width correction factor that also depends upon material anisotropy.

Using FE analysis, K_I for an orthotropic T800/924C carbon fibre/epoxy laminate, with cracks emanating symmetrically from a circular hole, is now determined.

A number of techniques are currently used for evaluating SIFs. Here, the Virtual Crack Extension (VCE) procedure in the FE77 finite element package, with an eight-node isoparametric element, is used to obtain the SIF for the problem shown in Fig. 8.5.

The fundamental difficulty in a crack problem is the singularity in the strain field at the crack tip. The finite element mesh must be such that the singularity is approximated with sufficient accuracy. Many methods have been devised to arrive at such an approximation; one commonly used method is the *quarter point element* where a degenerate form of the stan-

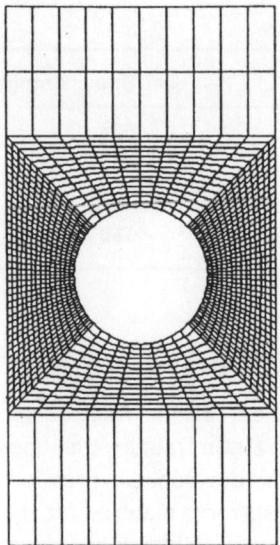

8.6 FE mesh around circular cut-out.

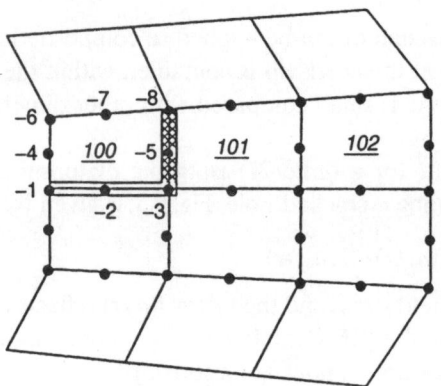

8.7 FE representation of crack.

dard eight-node quadrilateral element is employed. This isoparametric element produces the $r^{-1/2}$ singularity when it degenerates to a triangular element and the mid-side nodes are moved to the quarter point adjacent to the crack tip node.[13] In the FE77 code a refined mesh is required, Fig. 8.6, where the 'interface disconnect' subroutine is used to represent the crack, Fig. 8.7.

The FE77 program allows the evaluation of the J-integral by disconnecting nodes in certain elements ahead of the crack tip, Fig. 8.8, to represent a differential crack advance, δl, and measuring the change in strain

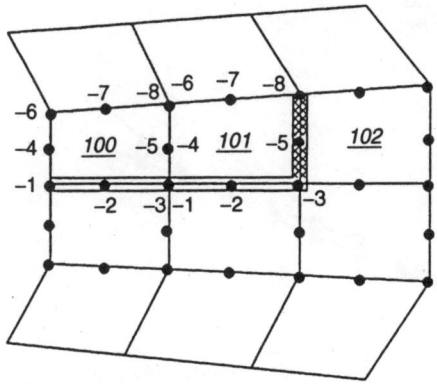

8.8 Incremental advance of crack.

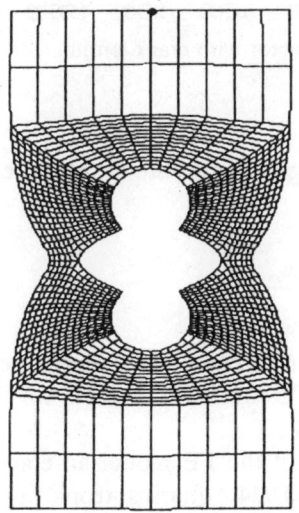

8.9 Deformed mesh for cracked hole.

energy δU. The J-integral is the negative differential of strain energy with respect to crack advance. For an elastic body J is identical to the elastic strain energy release rate G. Hence;

$$G = J = -\frac{\delta U}{\delta l} \qquad\qquad [8.11]$$

A typical deformed configuration of the plate with a cracked hole loaded in tension is shown in Fig. 8.9.

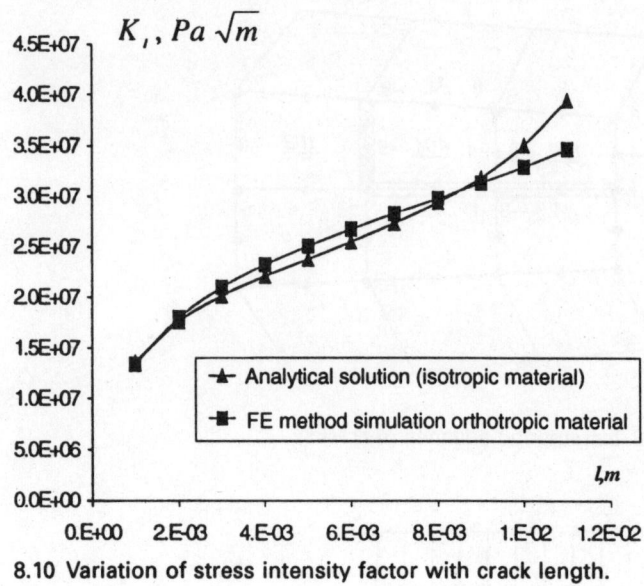

8.10 Variation of stress intensity factor with crack length.

For an orthotropic laminate the energy release rate G_I is related to the stress intensity factor K_I by:

$$G_I = \left(\frac{a_{11}a_{12}}{2}\right)^{1/2}\left[\left(\frac{a_{22}}{a_{11}}\right)^{1/2} + \frac{2a_{12}+a_{66}}{2a_{11}}\right]^{1/2} K_I^2 \qquad [8.12]$$

where a_{ij} are the coefficients of the laminate compliance matrix, related to elastic constants E_x, E_y, G_{xy} and υ_{xy}.[7] Of course, $G_I = K_I^2/E$ for isotropic materials.

In order to investigate the accuracy of the FE model an elastic stress analysis is performed. The SIF for a T800/924C quasi-isotropic $(\pm45/0/90°)_s$ cracked hole specimen, with $r/W = 0.25$, is shown in Fig. 8.10 and is compared with the analytical solution for an isotropic plate. It is clear from these results that the singularity in the stress field at the crack tip is approximated with sufficient accuracy, and the difference in the solution for K_I is small. Of course, the SIF cannot strictly be implemented in the inherent flaw fracture model described in Section 8.4.2 to predict the ultimate tensile strength of any orthotropic notched laminate, since the Bowie correction factor in Table 8.1 is for an isotropic plate.

For the quasi-isotropic $(\pm45/0/90°)_s$ laminate, however, the anisotropic effect is negligible and the isotropic SIF can be employed. The method described here is more useful for anisotropic lay-ups where an analytical solution does not exist.

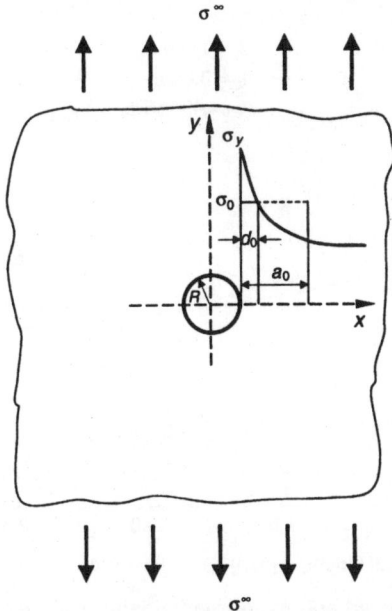

8.11 Point and average stress failure criterion.[9]

8.4.4   Point stress and average stress failure criteria

Whitney and Nuismer[9] have extended the Waddoups model[6] to explicitly include the anisotropy of the material. In their analysis, orthotropic symmetry is assumed. For a plate containing a circular hole of radius R, the ratio of notched strength to the unnotched strength is given by:

$$\frac{\sigma_n}{\sigma_{un}} = 2[2 + \zeta_1^2 + 3\zeta_1^4 - (K_T^\infty - 3)(5 - \zeta_1^6 - 7\zeta_1^8)]^{-1} \qquad [8.13]$$

where $\zeta = R/(R + d_0)$ and K_T^∞ = stress concentration factor for an orthotropic laminate.

As in the Waddoups' analysis, the assumption is that failure occurs when the stress at a point, a distance d_0 from the hole edge, reaches a critical value; Fig. 8.11. We can, in an alternative model, assume failure to occur when the average value of σ_y over a distance a_0 in front of the hole exceeds the unnotched tensile strength of the material; Fig. 8.11. In this case:

$$\frac{\sigma_n}{\sigma_{un}} = 2(1 - \zeta_2)[2 - \zeta_2^2 - \zeta_2^4 + (K_T^\infty - 3)(\zeta_2^6 - \zeta_2^8)]^{-1} \qquad [8.14]$$

where $\zeta_2 = R/(R + a_0)$.

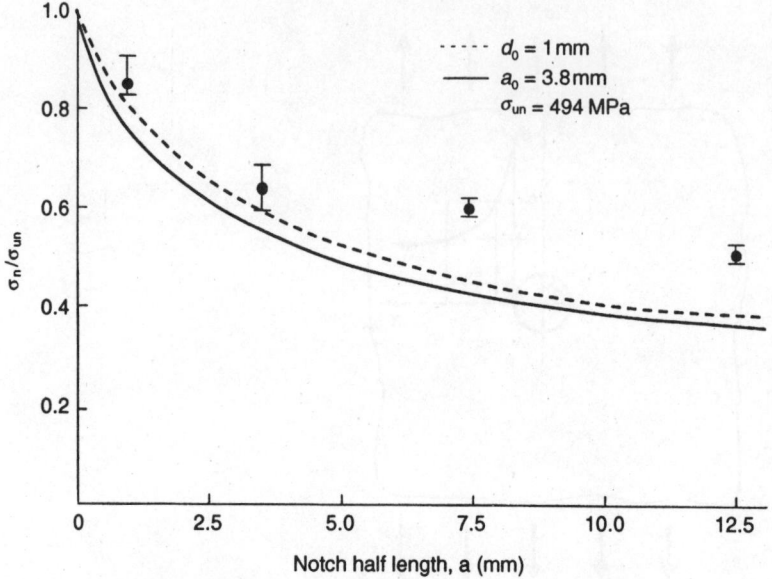

8.12 Comparison of predicted and experimental failure stresses for centre cracks in $(0/\pm45/90°)_{2s}$ HTS carbon fibre/epoxy resin.

For an infinite anisotropic plate containing a centre crack of length $2c$, subjected to a uniform uniaxial tensile load, the point stress failure criterion gives:

$$\frac{\sigma_n}{\sigma_{un}} = (1 - \zeta_3^2)^{\frac{1}{2}}$$ [8.15]

and the average stress failure criterion gives:

$$\frac{\sigma_n}{\sigma_{un}} = \left(\frac{1-\zeta_4}{1+\zeta_4}\right)^{\frac{1}{2}}$$ [8.16]

Here, $\zeta_3 = c/(c + d_0)$ and $\zeta_4 = c/(c + a_0)$.

In the Whitney–Nuismer models the characteristic length a_0 (or d_0) is used as a free parameter to be fixed by best fitting the experimental data. An example of a comparison between theory and experimental data is shown in Fig. 8.12 for centre notched $(0/\pm45/90°)_{2s}$ laminates made from high tensile strength (HTS) carbon fibre/epoxy. Here σ_n/σ_{un} is plotted against notch length c. The theoretical curves of σ_n/σ_{un} against notch size are computed using equations [8.15] and [8.16]. Values of $d_0 = 1$ mm and $a_0 = 3.8$ mm provided the best fit to these and other data for laminates with both circular holes and central cracks.

The models described above assume that K_{IC} and a_0 (or d_0) are invariant material properties. However, the fracture toughness and characteristic lengths depend on fibre and resin type, ply distribution and layer thickness. It is probably better to consider these parameters as *laminate* rather than *ply* properties, that have to be determined experimentally.

8.5   Progressive failure models

Recently, more interest has been placed on progressive failure modelling.[14-16] The mechanisms for damage progression and accumulation in notched laminates are so complicated (combinations of matrix cracking, longitudinal splitting, delamination and fibre breakage) that analytical methods are impractical and, perhaps, unable to model them. The most suitable tool is probably the FE method.

The damage progression around a hole can be determined by performing a ply-by-ply and element-by-element stress analysis and applying appropriate stress or strain-based failure criteria. If no laminae have failed, the load must be determined at which the first lamina fails, that is, violates the failure criterion. The load parameter is increased until some lamina fails. That lamina is then eliminated from the laminate by assigning zero stiffness (ply discount technique), or reduced properties (stiffness degradation factors), to the failed layer and new laminate extensional, coupling and bending stiffnesses (**ABD**) are calculated. Laminae stresses are recalculated to determine their distribution after a lamina has failed (the stresses will increase to maintain equilibrium). Then it must be verified that the remaining laminae, at their increased stress levels, do not fail at the same load that caused failure of the lamina in the preceding pass through the analysis. If no more laminae fail, then the load can be increased until another lamina fails and the cycle is repeated until last ply failure.

The overall procedure for strength analysis is independent of the failure criterion, but the results of the procedure, the maximum loads and deformations, will depend on the specific failure criterion. The laminate behaviour could be piecewise non-linear if the laminae behaved in a non-linear elastic manner. The results of such models include the whole field stress–strain relationship, the laminate stiffness degradations and the extent of damage at any stress level. The computing time varies significantly from one mesh design to another, depending on the number of elements. Although a finer mesh predicts failure progression better than does a coarse mesh, it needs more computing time. The increase in computing time may not be worth the increase in accuracy after the number for elements reaches a certain value. Figure 8.13(a) and (b) illustrate damage growth, in the form of matrix cracking, developed in the 90° ply of a (0/90°)$_s$ carbon fibre/epoxy laminate at uniaxial tensile loadings of 43.8 kN and 97.18 kN, respectively.

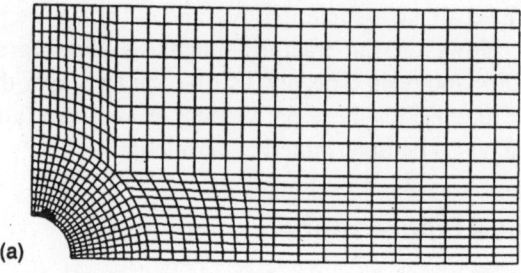

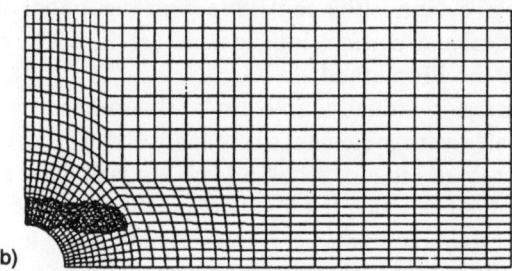

8.13 Finite element modelling of matrix cracking growth in the 90° ply
of a cracks in (0/90°)$_s$ CFRP laminate: (a) applied load of 43.8 kN;
(b) applied load of 97.18 kN.

8.6   Summary

We have seen that a number of parameters such as the critical strain energy
release rate, G_{Ic}, and the critical stress intensity factor, K_{Ic}, may specify the
toughness of a material. Fracture toughness parameters enable the critical
flaw size in a component subjected to a given stress to be determined.

Under some circumstances linear elastic fracture mechanics (LEFM)
can be used directly to describe fracture behaviour in composites and
measure their notch sensitivity. Typically this applies to 0° dominated
laminates of brittle composites (e.g. (0/±45°)$_s$, (0/90°)$_s$) with high fibre/resin
bond strengths and brittle matrices. In such laminates crack paths may be
planar, or nearly so, and damage zones small (e.g. 1–2 mm). LEFM seems
to predict notched strength reasonably well in quasi-isotropic (0/±45/90°)$_s$
laminates where there is a lower percentage of 0° fibres. The methods of
LEFM become invalid for angle-ply dominated laminates, e.g. (±45°)$_s$,
which are commonly notch insensitive. For these laminates the damage is
diffuse in nature, and a crack-like representation of damage becomes
inappropriate.

The important message is that generalisations should not be made. There is always likely to be uncertainty over the question of validity of LEFM for any given case.[1]

The FE method can be used successfully to predict the orthotropy effect on SIFs that are required in stress based failure criteria. It can also be used to model damage progression around open holes by performing ply-by-ply and element-by-element analysis, taking into account geometric and material non-linearities.

8.7   References

1 Harris B, *Engineering Composite Materials*, London, Institute of Materials, 1986.
2 Griffith A A, 'Fracture', *Phil Trans Roy Soc, London*, 1920 **Ser. A221** 163 (Republished with additional commentary in Trans ASM, 1968 **61**, 871).
3 Orowan E, *Fatigue and Fracture of Metals*, Cambridge, MA, MIT Press, 1950.
4 Irwin G R, *Fracturing of Metals*, Cleveland, OH, ASM, 1949.
5 Anderson T L, *Fracture Mechanics: Fundamentals and Applications*, Boca Raton, CRC Press, 1991.
6 Waddoups M E, Eiseman J R & Kaminski B E, 'Macroscopic fracture mechanics of advanced composite materials', *J Compos Mater*, 1971 **5**(10) 446–454.
7 Soutis C, Fleck N A & Smith P A, 'Failure prediction technique for compression loaded carbon fibre-epoxy laminate with open holes', *J Compos Mater*, 1991 **25**(11) 1476–1498.
8 Soutis C & Fleck N A, 'Static compression failure of carbon fibre T800/9246 composite plate with a single hole', *J Compos Mater*, 1990 **24**(5) 536–558.
9 Whitney J M & Nuismer R J, 'Stress fracture criteria for laminated composites containing stress concentrations', *J Compos Mater*, 1974 **8**(7) 253–265.
10 Rhodes M D, Mikulus M M & McGowan, 'Effects of orthotropy and width on the compression strength of graphite/epoxy panels with holes', *AIAA J*, 1984 **22**(9) 1283–1292.
11 Bowie O L, 'Analysis of an infinite plate containing radial cracks originating from the boundary of an internal circular hole', *J Maths Phys*, 1956 **35**(1) 60–71.
12 Paris P C & Sih G C, 'Stress analysis of cracks, fracture toughness testing and its implications', ASTM STP 381, Philadelphia, PA, American Society for Testing and Materials, 1965.
13 Marti V & Valliappon S, 'A universal optimum quarter point element', *Engineering Fracture Mechanics*, 1986 **25**(2) 237–258.
14 Tan S C, *Stress Concentrations in Laminated Composites*, Lancaster, PA, Technomic Publishing Co Inc, 1994.
15 Chang F K & Chang K Y, 'A progressive damage model for laminated composites containing stress concentrations', *J Compos Mater*, 1987 **21**(9) 834–855.
16 Tan S C & Nuismer R J, 'A theory for progressive matrix cracking in composite laminates', *J Compos Mater*, 1989 **23**(10) 1029–1047.

9.1 Introduction

The fracture processes in composite laminates under tensile monotonic and fatigue loading involve a sequential accumulation of damage in the form of matrix cracking, local delamination and edge delamination prior to catastrophic failure.[1-4] Edge delamination initiates at load-free edges of the composite plate due to Poisson's mismatch (see Chapter 7), whereas local delamination originates from the interaction of matrix ply cracks at ply interfaces.[5] These 'resin-dominated' failure modes can be detrimental to the strength of the laminate since they can cause fibre fracture in the primary load-bearing (0°) plies. It is therefore important to be in a position to predict their onset strain and growth.

Over the years a number of researchers have shown that the strain energy release rate, G, associated with local delamination can be compared with an appropriate value of interlaminar fracture toughness, obtained experimentally,[6] to estimate the initiation and propagation of delamination. O'Brien[3,4] derived a closed-form equation for G which, for certain lay-ups, can successfully predict the delamination onset strain. However, the author neglected the effect of matrix ply cracking which becomes important for laminates containing thick 90°-plies where local delaminations usually develop. More recently, Zhang and Soutis[7] modelled the matrix crack tip delamination and its effect on the stiffness properties of carbon fibre/epoxy laminates. A 2D shear lag approach was followed to obtain the microstress field in the damaged laminate. These resulting microstresses were used in an *in situ* damage effective function (IDEF) which was derived explicitly in terms of matrix crack density and relative local delamination area. Then, expressions for the strain energy release rates for matrix cracking and local delamination were derived, taking into account residual hygrothermal stresses and the interaction effect between the two damage modes. Further details are given below.

As mentioned in Chapter 6, FE analyses have used a number of

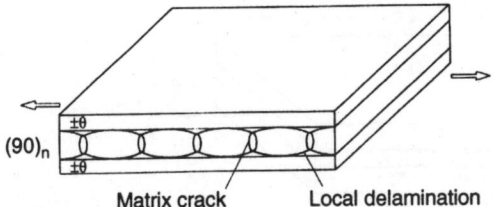

9.1 The $(\pm\theta_m/90°_n)_s$ laminate with matrix cracks and local delamination.

approaches to represent delamination, including the Virtual Crack Closure (VCC) method, as well as employing special 'delamination' elements (Sections 6.8 and 6.10). Again, results from such analyses will be given below.

9.2   A continuum model

When a $(\pm\theta_m/90°_n)_s$ balanced symmetric laminate is under static or fatigue tensile loading, matrix cracking in the transverse plies (90°) is the first damage mode observed and multiplies with increasing applied load or number of cycles. Subsequently, either edge delamination will form at the $-\theta/90°$ interface, or local delamination will initiate from the transverse ply crack tips due to high local stresses at the crack tip. A schematic of the damage configuration is shown in Fig. 9.1.

The mid-region is made of 90° plies, while the constraining layers (sublaminate) consist of multi-orientation plies $(\pm\theta)$. Transverse ply cracks are assumed to exist in the 90° plies with uniform crack spacing of $2s$; local delaminations initiate and grow from both tips of each transverse crack and span the width of the specimen. Following previous work,[7-9] in order to examine the effect of these damage modes on the laminate loading capacity, a representative three-layer segment, Fig. 9.2, containing a single transverse ply crack and two strip-shaped delaminations is considered. This element can be segregated into a locally delaminated region $0 \le y \le l_d$ and a non-delaminated region $l_d \le y \le s$. The effect of the cracking and the delaminations is to reduce the stiffness of the laminate.

9.2.1 Stiffness reduction

The modified stiffness matrix of a cracked lamina has been derived[8] by introducing the IDEF, Λ_{ij}, and is given by:

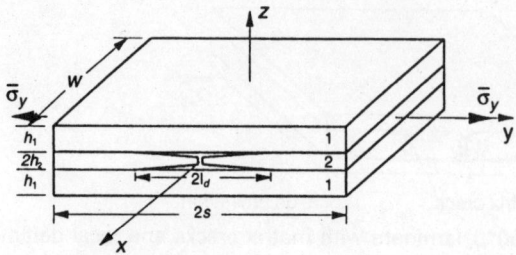

9.2 A three-layer element with local delamination growing from a 90°
matrix crack.

$$
\begin{bmatrix} Q_{11}^{(2)} & Q_{12}^{(2)} & 0 \\ Q_{12}^{(2)} & Q_{22}^{(2)} & 0 \\ 0 & 0 & Q_{33}^{(2)} \end{bmatrix} = \begin{bmatrix} Q_{11}^0 & Q_{12}^0 & 0 \\ Q_{12}^0 & Q_{22}^0 & 0 \\ 0 & 0 & Q_{33}^0 \end{bmatrix}
$$

$$
- \begin{bmatrix} \dfrac{(Q_{12}^0)^2}{Q_{22}^0}\Lambda_{22} & Q_{12}^0\Lambda_{22} & 0 \\ Q_{12}^0\Lambda_{22} & Q_{22}^0\Lambda_{22} & 0 \\ 0 & 0 & Q_{33}^0\Lambda_{33} \end{bmatrix} \qquad [9.1]
$$

where Q_{ij}^0 is the in-plane stiffness matrix of the 90° (ply 2) uncracked
lamina. The Λ_{22} and Λ_{33} parameters describe the stiffness loss of the 90° ply
caused by matrix cracking; they are functions of the average matrix crack
density, C_{d}, and also of the *in situ* constrained conditions of the 90° plies. In
order to calculate Λ_{22} and Λ_{33} a 2D shear-lag analysis was proposed,[7,8] where
the out-of-plane shear stresses varied linearly across the thickness of the
constraining layers (ply 1); Fig. 9.3.

Combining this shear stress variation with out-of-plane stress/strain and
strain/displacement relations, and taking the average values across the ply
thickness, the interface shear stresses were expressed in terms of in-plane
displacements. Once the microstress field in the cracked laminate was
obtained the macro stresses in the cracked 90° plies could be determined
by integrating over the laminate length. Substituting these stresses into the
damaged constitutive equation of the 90° plies, the Λ_{22} and Λ_{33} were derived
as functions of matrix crack density.

When delamination is present, growing from the transverse ply crack tips,
the locally delaminated portion of 90° plies is not able to carry normal load
in the y-direction and shear loads in xz and yz-planes; the overall laminate
stiffness properties are further reduced by the local delaminations. The
influence of the local delamination and matrix cracking on the overall laminate response was considered[5] by including the two damage modes in the

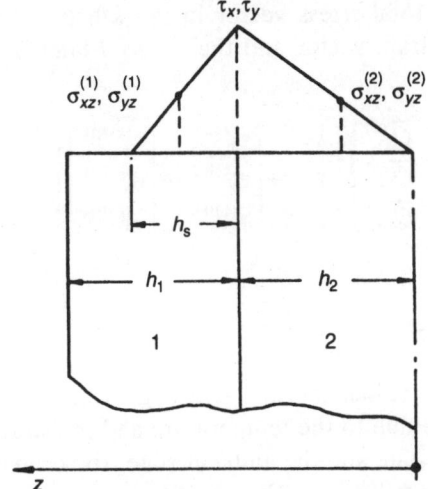

9.3 Out-of-plane shear stress variation across the thickness of the 90°
ply (ply '2') and the constraining layer (ply '1').

reduced lamina stiffness matrix of the 90° plies; equation [9.1]. The Λ_{22} and
Λ_{33} coefficients are given by

$$\Lambda_{22} = 1 - \frac{\Phi_1(1-D^{\mathrm{ld}}) + \Phi_2 D^{\mathrm{mc}} \tanh\left(\dfrac{\lambda_1(1-D^{\mathrm{ld}})}{D^{\mathrm{mc}}}\right)}{\Phi_1(1+\zeta_1 D^{\mathrm{ld}}) + \Phi_3 D^{\mathrm{mc}} \tanh\left(\dfrac{\lambda_1(1-D^{\mathrm{ld}})}{D^{\mathrm{mc}}}\right)} \qquad [9.2a]$$

$$\Lambda_{33} = 1 - \frac{\Gamma_1(1-D^{\mathrm{ld}}) + \Gamma_2 D^{\mathrm{mc}} \tanh\left(\dfrac{\lambda_2(1-D^{\mathrm{ld}})}{D^{\mathrm{mc}}}\right)}{\Gamma_1(1+\zeta_2 D^{\mathrm{ld}}) + \Gamma_3 D^{\mathrm{mc}} \tanh\left(\dfrac{\lambda_2(1-D^{\mathrm{ld}})}{D^{\mathrm{mc}}}\right)} \qquad [9.2b]$$

where $D^{\mathrm{mc}}(= h_2/s = 2h_2 C_d)$ and $D^{\mathrm{ld}}(= l_d/s = 2l_d C_d)$ are the relative matrix crack
density and the relative local delamination area, respectively. The parame-
ters $\Phi_i, \Gamma_i, \zeta_i$ and λ_i are functions of laminate elastic constants. When delamination is ignored, equations [9.2a] and [9.2b] reduce to the corresponding
relations for pure matrix cracking derived in reference 7.

9.2.2 Constitutive relationship of the damaged laminate

Let a laminate be subjected to in-plane loading under iso-hygrothermal
conditions, and be deformed and cracked progressively. Following previous

work[8,9] the macro-stress or total stress vector in the kth ply group can be written in terms of the strain vector and the reduced lamina stiffness properties

$$\begin{bmatrix} \sigma_x^{(k)} \\ \sigma_y^{(k)} \\ \sigma_{xy}^{(k)} \end{bmatrix} = \begin{bmatrix} \overline{Q}_{11}^{(k)} & \overline{Q}_{12}^{(k)} & \overline{Q}_{13}^{(k)} \\ \overline{Q}_{12}^{(k)} & \overline{Q}_{22}^{(k)} & \overline{Q}_{23}^{(k)} \\ \overline{Q}_{13}^{(k)} & \overline{Q}_{23}^{(k)} & \overline{Q}_{33}^{(k)} \end{bmatrix} \left(\begin{bmatrix} \varepsilon_x \\ \varepsilon_y \\ \gamma_{xy} \end{bmatrix} + \begin{bmatrix} \varepsilon_x^{(k)RT} \\ \varepsilon_y^{(k)RT} \\ \gamma_{xy}^{(k)RT} \end{bmatrix} + \begin{bmatrix} \varepsilon_x^{(k)RH} \\ \varepsilon_y^{(k)RH} \\ \gamma_{xy}^{(k)RH} \end{bmatrix} \right) \qquad [9.3a]$$

or

$$\sigma_k = \overline{Q}_k (\varepsilon + \varepsilon_k^{RT} + \varepsilon_k^{RH}) \qquad [9.3b]$$

where ε_k^{RT} and ε_k^{RH} are the residual thermal and hygroscopic (moisture) strain vectors in the laminate due to the temperature and moisture difference between the stress-free state and the ambient state. They can be computed using Classical Laminate Theory. The constitutive equation of the damaged laminate is found to be:

$$\begin{bmatrix} N_x \\ N_y \\ N_{xy} \end{bmatrix} = \begin{bmatrix} A_{11} & A_{12} & 0 \\ A_{12} & A_{22} & 0 \\ 0 & 0 & A_{33} \end{bmatrix} \left(\begin{bmatrix} \varepsilon_x \\ \varepsilon_y \\ \gamma_{xy} \end{bmatrix} - \begin{bmatrix} \varepsilon_x^p \\ \varepsilon_y^p \\ \gamma_{xy}^p \end{bmatrix} \right) \qquad [9.4a]$$

or

$$N = A(\varepsilon - \varepsilon^p) \qquad [9.4b]$$

with

$$\varepsilon^p = A^{-1}(N^T + N^H - A^{0-1}N^{0T} + N^{0H}) \qquad [9.5]$$

where A is the extensional stiffness matrix of the damaged laminate, and N^T and N^{0T} are the equivalent thermal loads in the damaged and undamaged states, with N^H and N^{0H} the respective equivalent moisture loads. It is observed that the laminate develops permanent strains due to the interaction effect between damage and residual hygrothermal stresses.

9.2.3 Energy release rate

The potential energy method of Zhang and Soutis[9] is used here to derive expressions for the energy release rates associated with local delamination, matrix cracking and their interaction. The potential energy (PE) of the equivalent damaged laminate element with a finite length $2l$ and width w is:

$$PE = U - N^t \varepsilon 2 l w \qquad [9.6]$$

where U is the total strain energy stored in the laminate element which is a function of matrix cracking density and local delamination area. Its calculation requires a ply-by-ply analysis in order to take into account the contribution of residual hygrothermal stresses. The energy release rate asso-' ciated with a particular damage mode is equal to the first partial derivative of the potential energy with respect to the crack surface area of the respective damage. The applied laminate loads, N, are fixed and the other damage modes remain unchanged. So, the strain energy release rates due to matrix cracking and local delamination are, respectively:

$$G^{mc} = -\left(\frac{\partial PE}{\partial A^{mc}}\right)_{\{N\},A^{ld}} \qquad [9.7a]$$

$$G^{ld} = -\left(\frac{\partial PE}{\partial A^{ld}}\right)_{\{N\},A^{mc}} \qquad [9.7b]$$

After some mathematical manipulation, it may be shown that, for specified stress resultants, the energy release rates for matrix cracking and local delamination are, respectively:

$$G^{mc}(N,D^{mc},D^{ld}) = -h_2\left(N^t A^{-1}\frac{\partial \overline{Q}_2}{\partial D^{mc}}A^{-1}N\right)$$

$$+ 2N^t A^{-1}\frac{\partial \overline{Q}_2}{\partial D^{mc}}(\varepsilon_k^{RT} + \varepsilon_k^{RH} + \varepsilon^P) \qquad [9.8a]$$

$$+ \left(\varepsilon_k^{RT^t} + \varepsilon_k^{RH^t} + \varepsilon^{P^t}\right)\frac{\partial \overline{Q}_2}{\partial D^{mc}}(\varepsilon_k^{RT} + \varepsilon_k^{RH} + \varepsilon^P)$$

$$G^{ld}(N,D^{mc},D^{ld}) = -\frac{h_2}{2}\left(N^t A^{-1}\frac{\partial \overline{Q}_2}{\partial D^{ld}}A^{-1}N\right)$$

$$+ 2N^t A^{-1}\frac{\partial \overline{Q}_2}{\partial D^{ld}}(\varepsilon_k^{RT} + \varepsilon_k^{RH} + \varepsilon^P) \qquad [9.8b]$$

$$+ \left(\varepsilon_k^{RT^t} + \varepsilon_k^{RH^t}\varepsilon^{P^t}\right)\frac{\partial \overline{Q}_2}{\partial D^{ld}}(\varepsilon_k^{RT} + \varepsilon_k^{RH} + \varepsilon^P)$$

The second and third terms on the right-hand side of equation [9.8] are due to residual hygrothermal stresses and their interaction with damage. These expressions for the strain energy release rate are general and simple to use. Knowing the in-plane stiffness matrix of the cracked lamina for a given matrix crack density, C_d, and relative local delamination area, D^{ld}, the energy release rate is directly evaluated from equations [9.8a, b] by differentiating the reduced stiffness matrix with respect to the corresponding damage variable. The effect of interaction between matrix cracking and local delamination is explicitly included in the expression of the *in situ* damage effective function, Λ_{ij}, which is used to obtain the reduced stiffness matrix, $\overline{Q}$.

9.3   Two-dimensional finite element analysis

9.3.1 Virtual crack closure approach

In this section a 2D FE analysis of the three-layer laminate considered above is performed to obtain the interlaminar stresses and the strain energy release rate associated with a local delamination starting from the transverse ply cracks. This will assess the analytical model presented in the previous section and O'Brien's simple equation for G.[4]

We assume the laminate to be subjected to a uniform axial tensile strain ε_y. Because of symmetry, only a quarter of the laminate, defined by $0 \leq y \geq s$ and $0 \leq z \leq h$, is considered. Transverse ply crack density, C_d, determines the length of the quarter, s ($s = 0.5C_d$; half crack spacing), while the relative local delamination area, D^{ld}, defines the interface crack length, l_d ($l_d = s\, D^{ld}$). The FE package ABAQUS[10] was employed to obtain the microstress field in the region of the crack tip. The quarter model is preliminarily divided into the 24 rectangles (4×6) shown in Fig. 9.4(a).

The rectangles are then subdivided into smaller regions using the eight-node quadrilateral plane strain element with 4-Gauss points. The detailed mesh of the OO'–AA' region is shown in Fig. 9.4(b).

Five divisions are imposed in the interval AB. Multipoint constraints are used at the perfectly bonded section of the θ/90° interface. Symmetry conditions are applied by imposing zero nodal displacement in the z-direction at $z = 0$ plane and in the y-direction at $y = 0$ plane. Load is applied by specifying uniform displacement in the y-direction at $y = s$ plane.

9.3.2 Evaluation of strain energy release rate

The energy release rate, G^{ld}, in the finite element model is evaluated by using the VCC technique,[11] as described in Section 6.9, and running the model twice. In the first run the model is loaded with nodes e, f and g, h held together (Fig. 9.5a) by employing the multipoint constraints (MPC) command in the ABAQUS package. The forces to hold these nodes together are evaluated. In the second run the model is loaded in the same manner but nodes e, f and g, h are allowed to move apart (Fig. 9.5b) and the relative displacements of the nodes recorded.

If F_{ef}^y and F_{ef}^z are the components of force in y- and z-directions required to hold nodes e and f together in the first run, respectively, and δ_{ef}^y and δ_{ef}^z are the relative displacement components between nodes e and f in the second run, the mode I and mode II strain energy release rate can be evaluated as

$$G_I^{ld} = \frac{1}{2\Delta l_d}(F_{ef}^z \delta_{ef}^z + F_{gh}^z \delta_{gh}^z)$$ [9.9a]

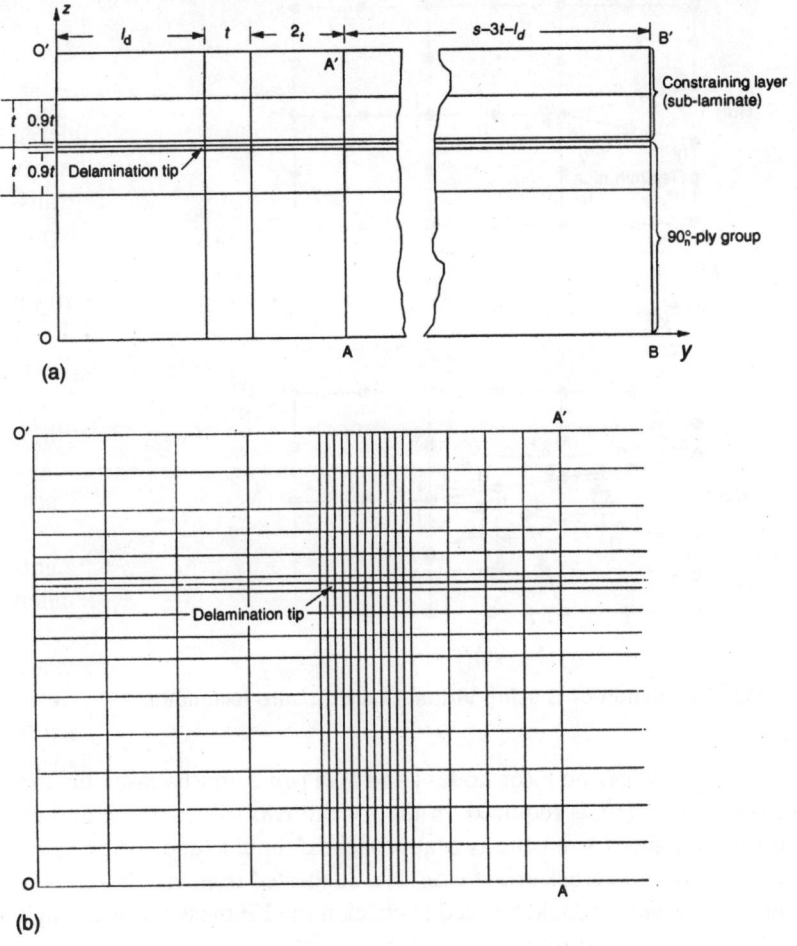

9.4 FE meshes for cracked-layer laminate: (a) coarse mesh and, (b) refined mesh at crack tip.

$$G_{II}^{ld} = \frac{1}{2\Delta l_d}\left(F_{ef}^y \delta_{ef}^y + F_{gh}^y \delta_{gh}^y\right)$$ [9.9b]

with the total energy being taken as:

$$G_T^{ld} = G_I^{ld} + G_{II}^{ld}$$ [9.10]

In equation [9.9] Δl_d is the increment of delamination growth (virtual crack extension).

Rybicki and Kanninen[11] suggested that the value of the nodal forces in equations [9.9a, b] could be replaced by the corresponding components of

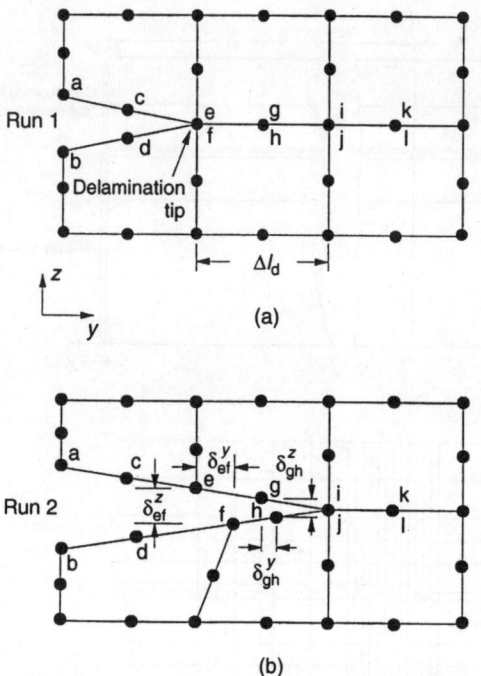

9.5 Calculation of *G* using Virtual Crack Closure technique.

nodal force of nodes i and k (or nodes j and l) in run 2. In this case only the second run, Fig. 9.5(b), is required. In theory, the two values of the energy release rate, obtained from the two approaches, should tend to the same value for small virtual crack growth, Δl_d. It is suggested that any discrepancy between the two values could be used to check if the FE mesh is fine enough to provide a reliable and stable value of the strain energy release rate.

9.3.3 An example

In order to compare the two virtual crack closure methods the nodal forces and the energy release rates are evaluated for a glass fibre/epoxy $(0_2/90_4^\circ)_s$ laminate. The lamina stiffness properties are:[12] $E_{11} = 50.33\,\text{GPa}$, $E_{22} = E_{33} = 14.48\,\text{GPa}$, $G_{12} = G_{13} = G_{23} = 6.068\,\text{GPa}$, $v_{12} = v_{13} = v_{23} = 0.275$ and the ply thickness $t = 0.254\,\text{mm}$. The transverse ply crack half-spacing, s, is assumed to be $10.16\,\text{mm}$ (= 40 ply-thicknesses); this corresponds to a matrix crack density $C_d = 0.492\,\text{cm}^{-1}$. The local delamination half-length is taken equal to $\Delta l_d = 1.016\,\text{mm}$ (= 4 ply-thicknesses), which is equivalent to relative delamination area $D^{ld} = 10\%$.

Figures 9.6 and 9.7 illustrate the variation of the interlaminar normal stress, σ_z, and shear stress, σ_{yz}, at the 0°/90° interface along the loading *y*-

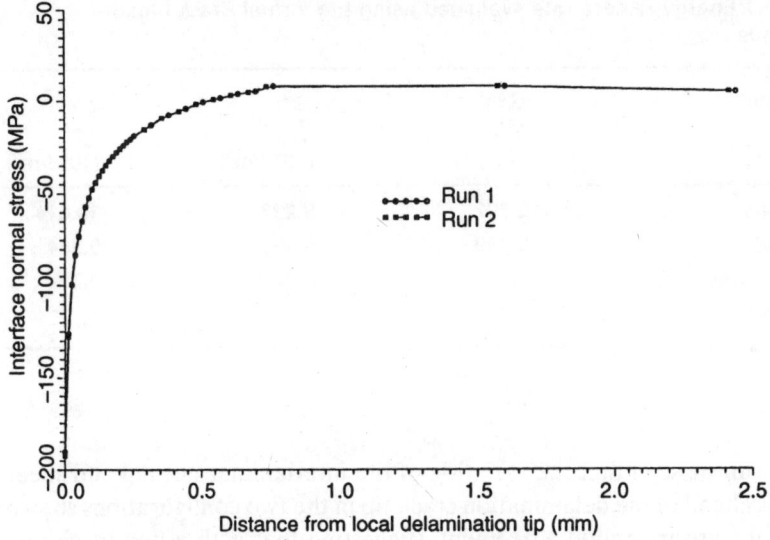

9.6 Variation of direct stress at the 0/90° interface of a $(0_2/90°_4)_s$ CFRP laminate.

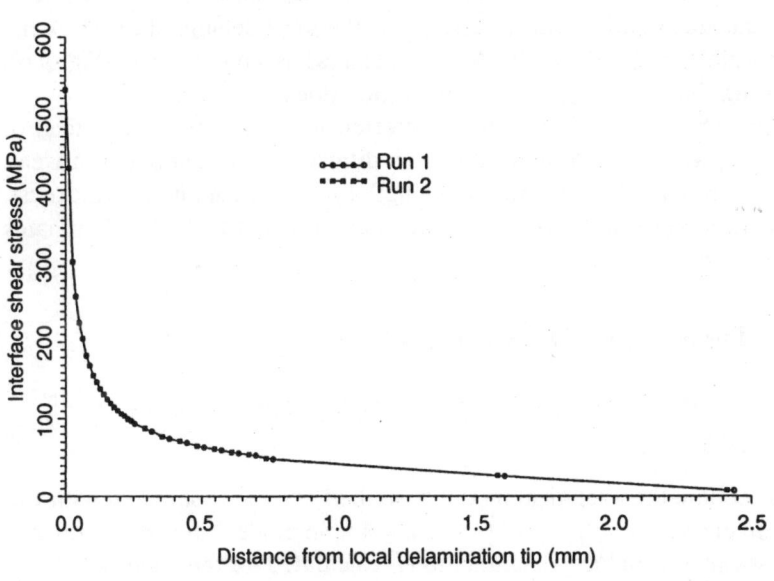

9.7 Variation of shear stress at the 0/90° interface of a $(0_2/90°_4)_s$ CFRP laminate.

Table 9.1 Energy release rate evaluated using the Virtual Crack Closure technique

FE models	$\dfrac{G_I^{ld}}{\overline{\varepsilon_y^2}}$ (10^6 J/m^2)	$\dfrac{G_{II}^{ld}}{\overline{\varepsilon_y^2}}$ (10^6 J/m^2)	$\dfrac{G_T^{ld}}{G_T^{ld}}$ (10^6 J/m^2)
Two runs	0.792	9.222	10.014
One run	0.799	9.195	9.994
Error between the two methods	0.9%	0.3%	0.2%

axis. It can be seen that the variation of the interlaminar normal and shear stresses ahead of the delamination crack tip in the two configurations shown in Fig. 9.5 are in a good agreement, suggesting that with a fine mesh one run can successfully evaluate the energy release rate due to local delamination. Results for G_T and its components are shown in Table 9.1; the maximum relative error between the two methods is only 0.9%.

Going back to the stress results, Fig. 9.6 and 9.7 show a sharp increase in interfacial shear and normal stresses near the local delamination crack tip. The normal stress, σ_z, at the 0°/90° interface is compressive (Fig. 9.6), suggesting that opening mode delamination does not exist.

In Fig. 9.8 the distribution of the interfacial normal stress, σ_z, ahead of the delamination tip is presented for four different local delamination areas. It can be seen that for delamination lengths greater than one-ply thickness, σ_z is compressive, indicating that the shear mode (mode II) dominates delamination growth.

9.4   Discussion of VCC results

9.4.1   Comparison between the analytical models and FE method

In this section, the analytical model described above is compared with the 2D finite element solution and O'Brien's simple model;[4] the energy release rate associated with local delamination is calculated for the T300/924 CFRP system with $(\pm 25/90°_4)_s$ lay-up. The FE analysis for local delamination is carried out for two different matrix crack densities; $C_d = 0.947 \text{ cm}^{-1}$ and $C_d = 4.38 \text{ cm}^{-1}$. The stiffness properties of the T300/924 system are taken as: $E_{11} = 144.8 \text{ GPa}$, $E_{22} = E_{33} = 11.38 \text{ GPa}$, $G_{12} = G_{13} = 6.48 \text{ GPa}$, $G_{23} = 3.45 \text{ GPa}$, $\upsilon_{12} = \upsilon_{13} = 0.3$ and ply thickness $t = 0.132 \text{ mm}$.

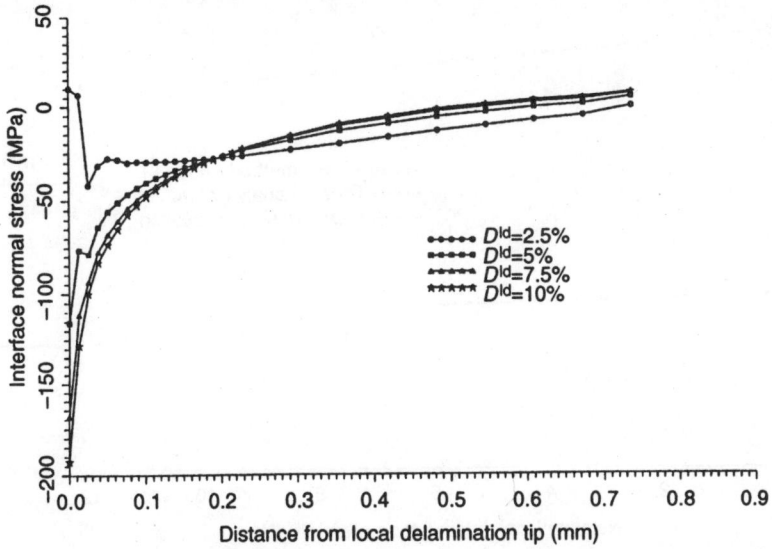

9.8 Variation of direct stress at the 0/90° interface of a (0₂/90°₄)ₛ CFRP laminate as a function of relative local delamination area, D^{ld}.

In Fig. 9.9 the energy release rate normalised by the applied strain is plotted against the relative local delamination area, D^{ld}, for the laminate with the two matrix crack densities. It can be seen that the strain energy release rate for delamination is reduced with increasing matrix crack density. The present analytical model agrees well with the FE solution in the steady-state growth of delamination. The predictions based on O'Brien's simple equation, which ignores the effect of matrix crack density, are far away from the FE results.

9.4.2 Effect of thermal residual stresses on G

In Fig. 9.10 the values of G with and without residual thermal stress effects are plotted against the relative delamination area. The applied axial strain is 1% and the matrix crack density $C_d = 4.38\,\text{cm}^{-1}$. The thermal coefficients parallel and perpendicular to the fibres are: $\alpha_1 = 0.36 \times 10^{-6}\,\text{C}^{-1}$ and $\alpha_2 = 28.8 \times 10^{-6}\,\text{C}^{-1}$.

The temperature difference between the stress-free state and ambient is assumed to be 125 °C and the specimen is assumed to be dry. It can be concluded that residual thermal stresses substantially increase the available energy to extend local delamination, and they should be included in predicting the local delamination onset strain (or stress).

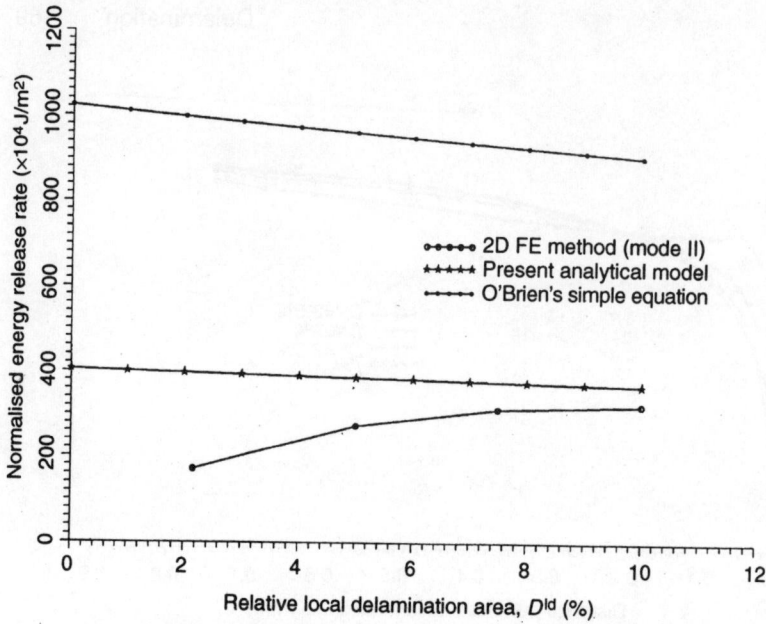

9.9 Variation of normalised energy release rate of local delamination with relative local delamination area, for the T300/914 graphite/epoxy $(\pm25/90°_4)_s$ laminate (matrix crack spacing, $80t$).

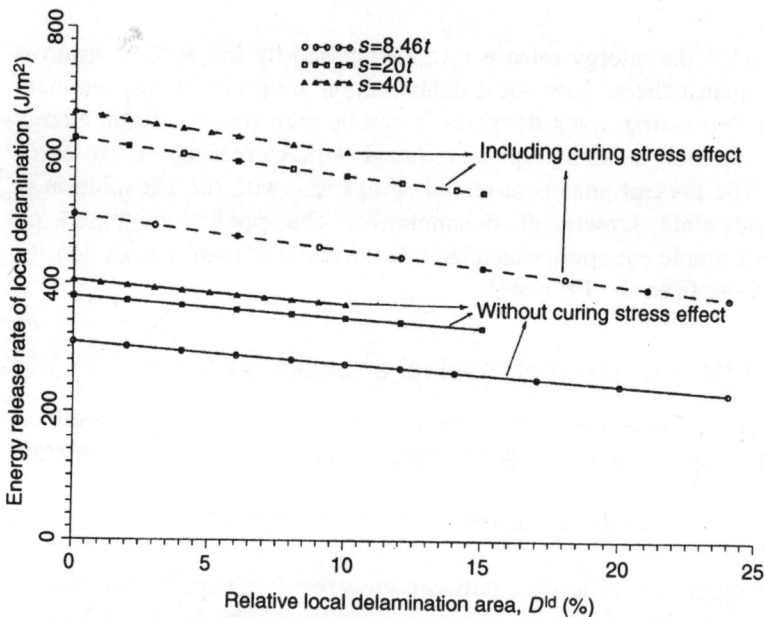

9.10 Variation of normalised energy release rate of local delamination with relative local delamination area, for the T300/914 graphite/epoxy $(\pm25/90°_4)_s$ laminate, with and without thermal stress (applied strain, 1%).

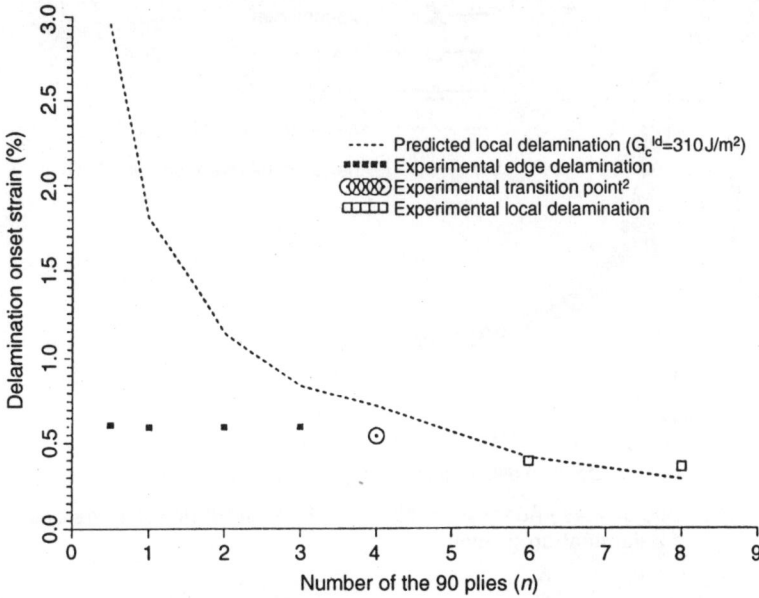

9.11 Delamination initiation strain for T300/914 graphite/epoxy
$(\pm 25/90°_n)_s$ laminates.

9.4.3 Local delamination initiation

In order to evaluate the critical load for local delamination initiation, the interlaminar fracture toughness, G_c^{ld}, of the composite material was taken as $310 J/m^2$. For the $(\pm 25/90°_n)_s$ laminates tested by Crossman and Wang,[2] the matrix crack density at which local delamination initiates can be inferred from the delamination onset strain and transverse ply crack density/strain data given in their paper. In Fig. 9.11 the dashed line gives the delamination onset strain as a function of the number of 90° plies in the laminate. In the experiments, the damage undergoes a transition from edge delamination ($n \leq 4$) to local delamination ($n \geq 4$). The prediction is in good agreement with experimental data for $n \geq 4$. For $n \leq 4$, the edge delamination is the dominant damage mode and has been analysed in Zhang et al.[13]

9.5 Finite element analysis using interface elements

The interface element described in Chapter 6 has been used to model several CFRP fracture mechanics test specimens.[14] To undertake the analyses, the linear interface element was implemented in ABAQUS and the

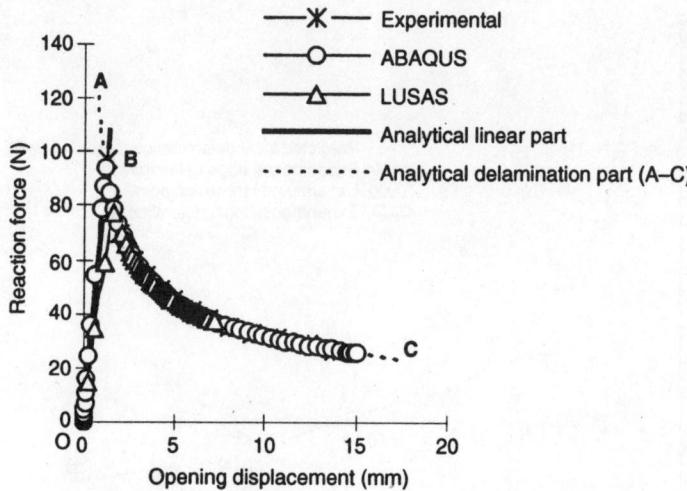

9.12 Load/displacement response for DCB. Point 'B' denotes the onset of delamination growth.[14]

quadratic element in LUSAS. The meshes used were such that there was roughly the same number of nodes in each case. The results indicated a faster convergence rate with the linear elements, although otherwise the two packages gave comparable results.

The situations modelled were the basic pure mode I test, the double cantilever beam (DCB), the pure mode II test, the end-loaded split (ELS), and the fixed ratio mixed mode (FRMM) test. The load–displacement response was computed and compared with experimental data, and with a closed form solution using modified beam theory.[15] As can be seen from Fig. 9.12, the comparison is excellent for the DCB specimen. The results for the ELS specimen were also good, although there was some difficulty in getting, simultaneously, a close agreement for both the initial (linear) and 'propagating' parts of the load–displacement response. Similar, but not such pronounced, differences were encountered with the FRMM model. These issuse could be overcome by modifying the values of E_{11} input to the analysis, but such an approach is not satisfactory and indicates that more work is required in developing these elements.

9.6   Summary

The modelling of cracking in laminates has been illustrated with reference to laminates with $(\pm\theta_m/90_n^\circ)_s$ lay-ups. A new theoretical model and a 2D finite element analysis were used as a basis for comparison. The total G_T

was seen to consist primarily of the G_{II} component, and the FE result using the VCC method was in good agreement with theoretical predictions for large delaminations. It was substantially affected by the matrix crack density and the residual thermal stresses.

The recently developed interface element has been shown to offer considerable promise, despite much development being needed, especially for application in complex structures.

9.7   References

1 Jamison R D, Schulte K, Reifsnider K L & Stinchcomb W W, 'Characterization and analysis of damage mechanisms in tension-tension fatigue of graphite/epoxy laminates', ASTM STP-836, Philadelphia, PA, American Society for Testing and Materials, 1984.

2 Crossman F W & Wang A S D, 'The dependence of transverse cracking and delamination on ply thickness in graphite/epoxy laminates', ASTM STP-775, Philadelphia, PA, American Society for Testing and Materials, 1982.

3 O'Brien T K, 'Characterization of delamination onset and growth in a composite laminate', ASTM STP-775, Philadelphia, PA, American Society for Testing and Materials, 1982.

4 O'Brien K T, 'Analysis of local delaminations and their influence on composite laminate behaviour', ASTM STP-876, Philadelphia, PA, American Society for Testing and Materials, 1985.

5 Zhang J, Soutis C & Fan J, Strain energy release rate associated with local delamination in cracked composite laminates, *Composites*, 1994 **25**(9) 851–862.

6 Martin R H & Murri G B, 'Characterization of Mode I and Mode II delamination growth and thresholds in graphite/PEEK composites', ASTM STP-1059, Philadelphia, PA, American Society for Testing and Materials, 1990.

7 Zhang J, Fan J & Soutis C, 'Progressive matrix cracking in composite laminates loaded in tension', *Adv Composites Lett*, 1992 **1** (1) 12–15.

8 Zhang J, Fan J & Soutis C, 'Analysis of multiple matrix cracking in $[\pm\theta_m/90_n]_s$ composite laminates: Part I, In-plane stiffness properties', *Composites*, 1992 **23**(5) 291–297.

9 Zhang J, Fan J & Soutis C, 'Analysis of multiple matrix cracking in $[\pm\theta_m/90_n]_s$ composite laminates: Part II, Development of transverse ply cracks', *Composites*, 1992 **23**(5) 299–304.

10 *ABAQUS User's Manual*, Hibbitt, Karlesson & Sorensen, Inc, Providence, RI, 1992.

11 Rybicki E F & Kanninen M F, 'A finite element calculation of stress intensity factors by a modified crack closure integral' *Engineering Fracture Mechanics*, 1977 **9**(4) 931–938.

12 Salpekar S A & O'Brien K T, 'Combined effect of matrix cracking and free edge on delamination', ASTM STP-1110, Philadelphia, PA, American Society for Testing and Materials, 1991.

13 Zhang J, Soutis C & Fan J, 'Effects of matrix cracking and hygrothermal stresses on the strain energy release rate for edge delamination in composite laminates', *Composites*, 1994 **25**(1) 27–35.

14 Chen J, Crisfield M A, Kinloch A J, Matthews F L, Busso E & Qui Y 'Predicting delamination of composite material specimens via interface elements', *J Mech Composite Mater Structures*, 1999 6(1) 1–17.
15 Kinloch A J, Wang Y, Williams J G & Yayla P, 'The mixed-mode delamination of fibre composite materials', *Compos Sci Tech*, 1993 47(3) 225–237.

10
Joining

10.1 Introduction

Joints in components or structures incur a weight penalty, are a source of
failure and cause manufacturing problems; whenever possible, therefore,
a designer will avoid using them. Unfortunately it is rarely possible to
produce a construction without joints owing to limitations on material size,
convenience in manufacture or transportation and the need for access in
order to inspect or repair the structure.[1] This chapter describes various fastening methods commonly employed with composite materials, the types of
joint failure, and the kind of problems that arise in the joint design because
of the heterogeneous and anisotropic nature of composite materials.

Basically, there are two types of joint commonly employed with composite materials: adhesively bonded joints and mechanically fastened joints.
Welding is also a possibility for thermoplastic composites, but this technique
is not well developed for load-carrying joints. In the following paragraphs,
the first two types of joint are discussed.

10.2 Adhesively bonded joints

10.2.1 Introduction

Bonded joints can be made by bonding (glueing) together pre-cured laminates with a suitable adhesive or by forming joints during the manufacturing process, in which case the joint and the laminate are cured at the same
time (co-cured or co-bonded). Load-carrying joints often have an overlap
configuration; see Fig. 10.1. Such joint forms would be appropriate for
joining flat laminates or tubular members. Because the strength of a joint
is sensitive to a large number of parameters it is vital that consideration is
given to the joints at the outset of any design. Failure to do so may result
in the component being otherwise of adequate performance but impossible to join. In the development of bonded joints for structures, a simple

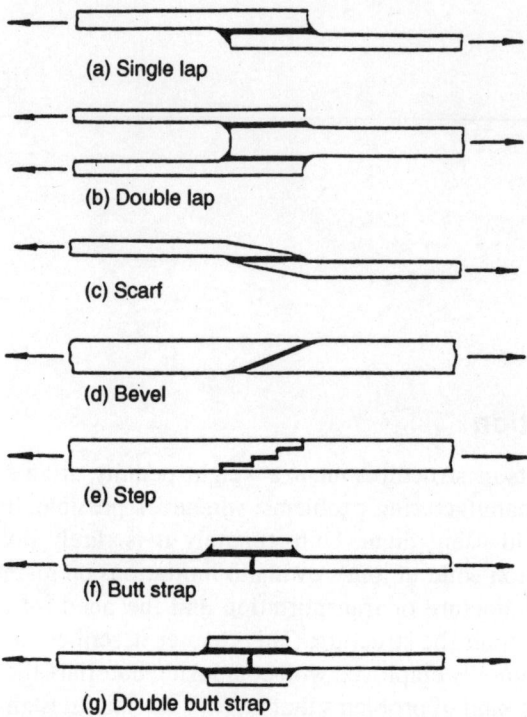

(a) Single lap

(b) Double lap

(c) Scarf

(d) Bevel

(e) Step

(f) Butt strap

(g) Double butt strap

10.1 Some common engineering adhesive joints.[1]

joint can be fabricated first and tested for its suitability in structures. The size of the joint can be estimated from a knowledge of the part sizes to be joined, the allotted space for the joint, and a general idea of how much overlap is required to carry the load. With such knowledge, preliminary joint designs can be made that can be refined using an iterative analysis procedure.

10.2.2 Stress distributions

A joint represents a discontinuity in a structure, and the resulting high stresses often initiate failure. Therefore, knowledge of the stresses in joints is vital if we are to understand the failures that occur in practice and hence improve designs and predict strength. Even relatively simple theories can be useful if they allow the important parameters to be identified. There are many publications concerned with the stress analysis of bonded joints,[2–11] and analyses have been carried out for various joint configurations and for different properties of the adherends and adhesives. Results have been obtained in closed form or from numerical analyses. Important results are

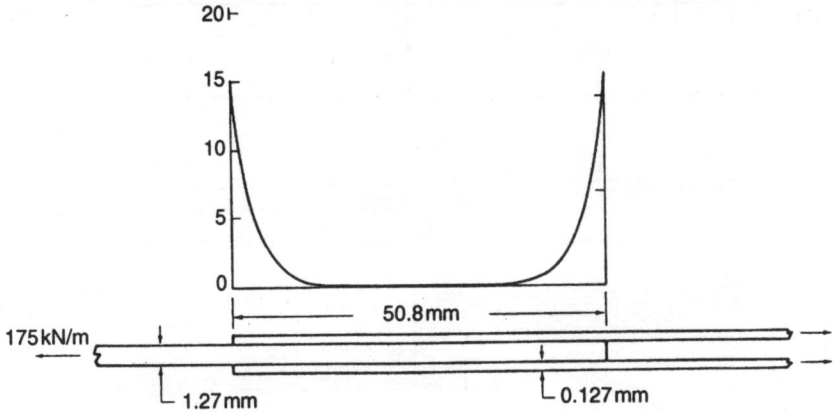

10.2 Adhesive shear stress distribution in bonded joints.[2]

qualitatively described here, and the conclusions affecting the joint design are also discussed.

The primary purpose of a joint is to transfer load between the two items being joined. In most bonded joints the load transfer takes place through interfacial shear, which gives rise to high interlaminar stresses in the adhesive layer. An idealised (elastic) variation of interlaminar shear stress in a double lap joint is shown in Fig. 10.2. It can be observed in the figure that the interlaminar shear stress has a large concentration near the end of the joint. In the remainder of the joint the stress is low and uniformly distributed. Because the load transfer zones occur at the ends and eventually reach a constant length, no increase in joint strength will be achieved once these zones are fully developed. There is no point in increasing the overlap beyond a critical value since no significant enhancement in strength will result. Typically, for carbon fibre/epoxy systems the limiting overlap is $30t_a$, where t_a is the thickness of the adherend. Since high stresses are developed in the adhesive layer at the ends of the overlap, high stresses are produced in the adjacent plies of the adherends. Therefore, failure may initiate in these plies. Hart-Smith[12] suggested that an effective way of reducing the local high stresses in the plies adjacent to the adhesive layer is to increase the adhesive thickness at the edge of the overlap. However, it is important to remember that good adhesive bonds can be produced only in small range of thicknesses (typically 0.1–0.25 mm) since thick bonds tend to be porous and weak while ultrathin bonds are too stiff and brittle.

In contrast to double lap joints, the shear characteristics of the adhesive have little influence on the strength of single lap joints, which are determined mainly by adherend properties, peel stresses and overlap length. The

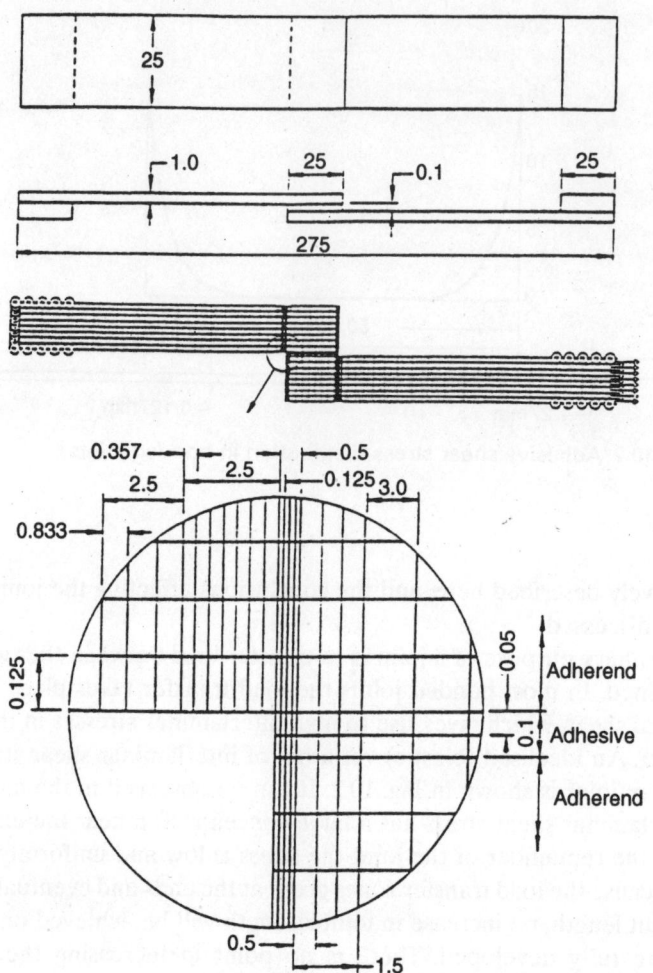

10.3 Typical single lap joint geometry and FE model (dimensions in mm).[13]

length of the plastic zone in the adhesive is only one half that in equivalent double lap joint and the elastic trough carries a significant proportion of the total load. Peel stresses are an order of magnitude greater than in a double lap joint and, in contrast to the latter situation, their effect can be minimised by increasing the lap length. For acceptable efficiency the overlap should be at least $80t_a$.

A typical FE mesh for a single lap joint between eight-ply laminates is shown in Fig. 10.3. There is one element through the depth of each ply, and two through the thickness of the adhesive. Notice the increase in mesh density at the ends of the overlap.[13] Continuum and FE analyses of such joints give the same results for comparable situations.[14]

Scarf joints (Fig. 10.1) between identical adherends will have a uniform distribution of adhesive shear stress and hence will show a higher strength than the other joint types. With non-identical adherends such stress uniformity is not obtained and the scarf joint will usually fail at the tip of the stiffer adherend. The same situation can occur with stepped joints. To restrict creep deformation the scarf angle must be kept small (1–3°), resulting in long joints. Stress analysis of both stepped and scarf joints between dissimilar adherends (e.g. composite and metal) must include the influence of the different adherend stiffnesses and thermal expansion coefficients.

10.2.3 Failure modes and strength

Micromechanical damage occurs first and will eventually lead to macromechanical damage. Thus micromechanical damage can be the basis for the selection of ultimate load prediction techniques and the prediction of failure modes of the joints. The micromechanical damage may initiate in the adhesive layer, at the adhesive/adherend interface, or in the adherends. Adherend failure can be tensile, interlaminar or transverse, in the last two cases either in the resin or at the fibre/resin interface.

Cohesive failure within the adhesive layer or in the surface layer of the adherend matrix may occur by brittle fracture or by a rubbery tearing, depending on the type of adhesive used. This results in cracking perpendicular to the load and causes a reduction in the load–transfer capability of the joint. This situation is analogous to the cracks in the 90° plies of a crossply laminate. Adhesive/adherend interface failure usually occurs on a macro scale at low loads when processing (surface treatment) or material quality are poor; it should not take place in properly prepared joints. Interlaminar failure in the laminate (not related to edge effects) may be caused by poor processing, voids, delaminations or thermal stresses.

The weakest joints are those where failure is limited by interlaminar failure of the adherend or peel of the adhesive. The next strongest joints are those in which the load is limited by the shear strength of the adhesive, while the strongest will fail outside the joint area at a load equivalent to the strength of the adherend. Such failures can be related to the stress distributions described earlier. Further details on the different types of joint failure, joint analysis procedures and joint design allowables can be found in Hart-Smith.[12]

10.2.4 FE analysis of scarf joint

In this section we consider a joint that has identical adherends, uses a relatively brittle adhesive and has small scarf angles, Fig. 10.4. The FE77

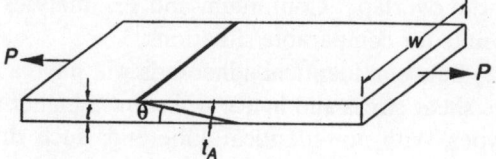

10.4 Scarf joint geometry.

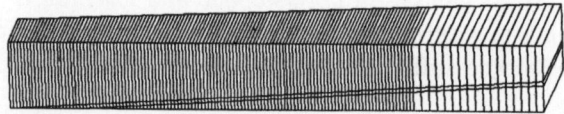

10.5 Refined FE mesh near scarf tip.

finite element package[15] is employed to determine the stress field in the scarf joint, using an isoparametric eight-node solid element. Because a high stress concentration is expected near the sharp tip, a fine mesh is required in this area; Fig. 10.5. The smallest element dimension is 0.0125 mm.[16] The $(\pm45/0/90°)_{2s}$ carbon fibre/epoxy laminate is treated as an homogeneous elastic material with the following stiffness/strength properties: $E_{xx} = E_{yy} = 53.8$ GPa, $E_{zz} = 11.3$ GPa, $G_{xy} = 20.5$ GPa, $G_{xz} = G_{yz} = 4.85$ GPa, $\upsilon_{xy} = 0.31$, $\upsilon_{xz} = \upsilon_{yz} = 0.19$ and compressive strength $\sigma_{un} = 454$ MPa. The epoxy adhesive layer of thickness $t_A = 0.129$ mm has the following properties: $E = 3.40$ GPa, $G = 1.26$ GPa, $\upsilon = 0.35$ and $\tau_s = 40$ MPa.

10.2.4.1 Stress results

It is found that the dominant stress components are the in-plane stress, σ_x, along the x-axis (load axis) in the parent laminate, and the shear stress tangential to the tapered bond surface. Other stress components are relatively small and could be neglected in any failure load calculations. The axial stress distribution has a steep gradient near the tip of the scarf, suggesting the existence of a stress singularity, and approaches the applied remote stress, S_g, within eight plies (1 mm); Fig. 10.6. Since a mathematical stress singularity exists at the scarf tip due to stiffness discontinuity, the computed stresses are taken near, but not at, the tip of the scarf. The elastic stress concentration factor, at a distance of 0.01 mm from the scarf tip, for the laminate examined, is approximately 2.3. As the distance from the tip increases, the in-plane direct stress rapidly decreases.

The shear stress in the adhesive is quite uniformly distributed; Fig. 10.7. Its magnitude is nearly proportional to the scarf angle; for $\theta = 9°$, it is almost three times higher than the value obtained for $\theta = 3°$.

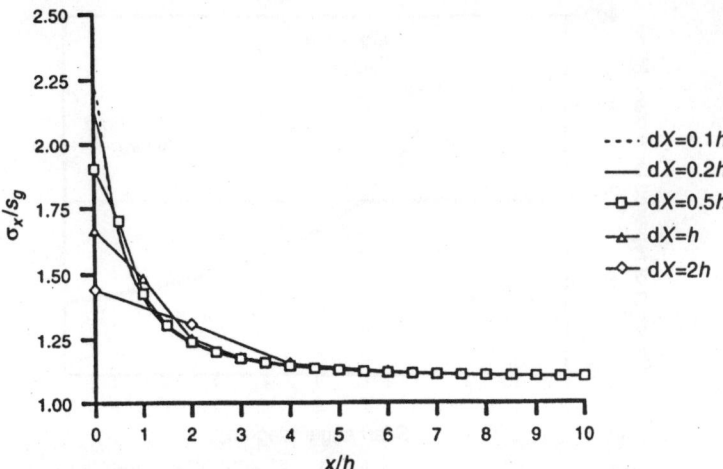

10.6 Distributions of σ_x/S_g in the parent plate along x-axis from the scarf tip; S_g is the remote stress.

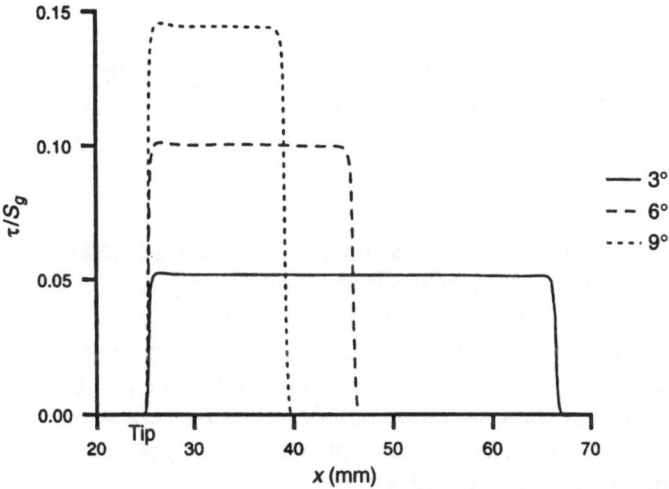

10.7 Distributions of normalised shear stresses, τ/S_g, in the adhesive layer along x-axis when the scarf angle equals 3, 6 and 9°.

10.2.4.2 Scarf joint strength and optimum scarf angle

For the simple scarf joint case, a simple stress analysis[16] predicts that the optimum scarf angle θ_{opt} for a maximum strength joint is a function of the adhesive shear strength τ_s and laminate strength σ_{un}, given by

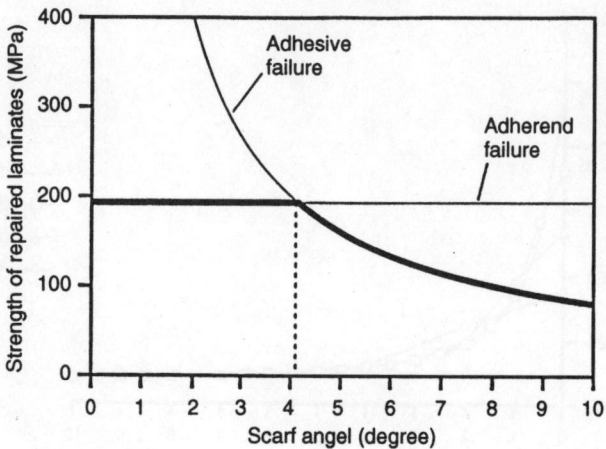

10.8 Failure stress variation with scarf angle based on maximum stress criterion.

$$\theta_{opt} \cong \tan^{-1}\left(\frac{\tau_s}{\sigma_{un}}\right) \qquad [10.1]$$

For small θ the failure stress, S_{gf}, of the scarf joint is determined by the maximum stress failure criterion, and is given by

$$S_{gf} = \frac{\sigma_{un}}{K_L} = \frac{\tau_s}{K_A \sin\theta} \qquad [10.2]$$

where K_A and K_L are the stress concentration factors (SCFs) in the adhesive and adherend, respectively. From the FE analysis, $K_A \cong 2.88$ and $K_L \cong 2.3$ for $\theta \leq 10°$. The load-carrying capability of the adhesive and the adherend is plotted in Fig. 10.8 as a function of scarf angle. The optimum scarf angle occurs when the adhesive failure load is equal to the laminate failure load. For the composite system examined ($\tau_s = 40\,MPa$ and $\sigma_{un} = 454\,MPa$), equation [10.2] results in a failure stress of $\cong 200\,MPa$ and an optimum scarf angle θ_{opt} of 4°.

Since stress redistribution, owing to material non-linearities and resin plasticity, may occur in the scarf region before final failure, using elastic SCFs could substantially underestimate the failure load. An alternative approach is that of averaging the stresses over a distance from the tip of the scarf, suggesting that the exact values of the stresses at the tip are not too important, see Section 8.4.4. This approach accounts for material non-linearities and plasticity of the adhesive, which reduce local peak stresses in the scarf region, and stress redistribution mechanisms, which are not considered in the elastic FE analysis.

10.3 Mechanically fastened joints

10.3.1 Introduction

In structures where parts are removed for inspection or maintenance bolted joints will be required. The behaviour of bolted connections, for composite laminates made from unidirectional pre-preg material, has been extensively examined by several workers,[17-29] who investigated a wide range of variables such as lay-up, fastener type (screw, rivet, bolt), friction effects, clearance and their influence on the failure mode. A full theoretical description of the stresses in such a joint must include their three-dimensional nature, a fact that has limited the analytical treatment given to such connections. The prediction of failure loads is, at the moment, mostly done semi-empirically. Improvement will depend on the development of failure criteria that are more generally applicable, together with an easy-to-use 3D stress analysis. In the latter context, FE analysis is clearly important, and some recent work is discussed below.

10.3.2 Failure modes

In addition to fastener failure, in shear and/or bending, there are essentially four modes of failure, namely tension, shear, bearing and cleavage, as illustrated in Fig. 10.9. Failure will be dependent upon many factors, such as fibre type, fibre orientation, surface treatment and matrix properties. It follows that knowledge of a wide range of variables is needed if favourable joint conditions are to be achieved and unwanted failure modes avoided. Composite materials may have low bearing strength and low in-plane shear strength. The tensile strength of the material in the reduced cross-section can be improved by increasing the spacing between bolts and transferring the load through several rows of bolts so that the net shear area of the bolts is sufficient.

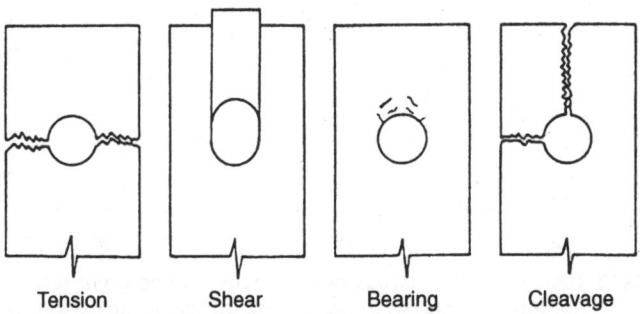

| Tension | Shear | Bearing | Cleavage |

10.9 Modes of failure for mechanical joints in FRP composites.

The low in-plane shear strength of the composite presents quite a few problems. Unidirectional composites have low shear strength in the longitudinal direction, which results in the shear out mode of joint failure. An improvement for this mode of failure can be made by the use of ±45° fibre orientations, but this can result in low net tension capability. The use of quasi-isotropic fibre orientations combines improved shear and tensile strength, but at the same time reduces considerably the efficiency of the joint.

Besides the preceding problems related to conventional strength criteria, there are problems peculiar to composite materials. The holes in the laminate cause stress concentrations that vary with the fibre orientation relative to the load direction. The stress concentration factors may sometimes be well above those occurring in a similar metal structure. Composite materials do not plastically deform, so that stresses are not easily redistributed around the stress concentration and are thus a cause of concern. The holes also give rise to the edge effects, as discussed in Chapter 7, which promote local interlaminar failures that may become critical in the presence of a corrosive environment. A detailed discussion of current joint design practices can be found in other works.[29,30]

10.3.3 Bearing strength prediction

There are three significant in-plane stresses in the laminate: the compressive (bearing) stress on the loaded side of the pin, the tensile stress across the net section and the shear stress on the shear out planes; Fig. 10.10.

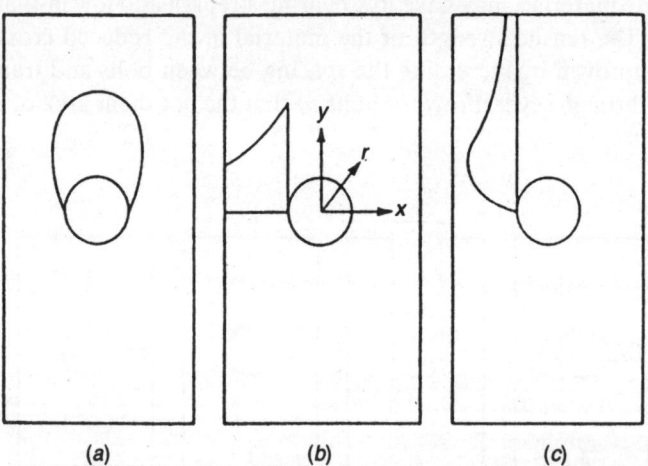

10.10 Typical in-plane stress distributions around a pin-loaded hole: (a) radial compression stress, σ_r; (b) tensile stress on net section, σ_y; and (c) shear stress on shear out planes, σ_{xy}.

Table 10.1 Measured and predicted bearing strengths of woven CFRP laminates

Laminate	Failure load, P_f (kN)	σ_b	Strengths (MPa) σ_{th}, Eqn [10.4]	σ_{th}[18]
$(0/90°)_{8s}$	12.6	553	315 (567)	620
$(0/90/± 45°)_{4s}$	16.5	723	363 (653)	402

Note: number in parentheses are strengths obtained by reducing the SCF.[29]

Provided certain geometric requirements are met joint failure will be in bearing.

Smith[24] and Smith *et al.*[26] related the maximum tensile stress concentration factor (SCF) at a loaded hole, K_b, to the maximum stress concentration factor at an unloaded hole in a finite width orthotropic laminate under a remote tensile stress by the following expression:

$$K_b = K_T^\infty [0.31 + 0.5Y(d/w)] \qquad [10.3]$$

K_b is defined as the maximum tensile stress normalised by the nominal bearing stress, Y is a finite width correction factor, d is the hole diameter and w is the plate width. K_T^∞ is related to the laminate elastic constants (E, v, G) and is given by Lekhnitskii.[31] For the woven carbon fibre (T300 2 × 2 twill)/epoxy (135 °C curing toughened epoxy, LTM49) $(0/90°)_{8s}$ orthotropic and $(0/90/±45°)_{4s}$ quasi-isotropic laminates investigated by Andreasson *et al.*,[32] K_b is equal to 2.063 and 1.188, respectively. These values could be used with the maximum stress criterion to predict the bearing strength (in the net tension mode of failure); i.e.

$$\sigma_{th} = \frac{\sigma_{ult}}{K_b} \qquad [10.4]$$

where σ_{ult} is the ultimate tensile strength of the laminate. Equation [10.4] predicts that failure would occur at 315 MPa for the $(0/90°)$ lay-up and at 363 MPa for the $(0/90/±45°)$ laminate; Table 10.1. Both values are much lower than the measured strengths, σ_b.

This suggests that some load redistribution is occurring and a simple stress concentration factor underestimates considerably the notched strength. A simple means of incorporating the load redistribution is to apply the point stress or average stress failure criterion,[33] or just reduce the elastic SCF by a constant C ($\cong 1.8$; determined from the best fit to experimental data) as suggested by Hart-Smith.[29]

Note that the parameter C for the non-woven laminates (XAS/914 carbon fibre/epoxy system) examined by Smith[24] and Smith et al.[26] was taken as 2.1, independent of stacking sequence and thickness.

In order to predict the bearing strength, in the bearing mode of failure, the semi-empirical technique developed by Collings[18] could be used. The method underestimates the bearing strength of the quasi-isotropic laminate by more than 40%, but in general the agreement is acceptable considering the complex nature of damage initiation and growth observed in the experiments.[32] It would be more appropriate to use Collings' expressions to estimate damage initiation rather than final fracture. The average or point stress criterion could be applied to account for the local damage occurring around the hole, but knowledge of the stress distribution is required. This can be achieved by performing a 2D FE stress analysis.

10.3.4 2D FE analysis

The stress field in the region of a loaded hole in a composite laminate is quite complicated, even if the problem is treated as being two-dimensional (load transfer friction and through-thickness clamping effects ignored). The stress distributions are influenced by the elastic constants of the laminate, the joint width and end distance, pin elasticity, friction and clearance between the pin and the hole. Andreasson et al.[32] used the commercial FE package I-DEAS to model a pin joint (i.e. a bolt with no through-thickness restraint or tightening) as a 2D contact problem which incorporated the bolt/hole friction and an appropriate contact region between the bolt and the plate. Orthotropic $(0/90°)_{8s}$ and quasi-isotropic $(0/90/\pm45°)_{4s}$ laminates were examined. For linear static problems I-DEAS makes use of laminate plate theory to determine ply and laminate stresses. A typical FE mesh is shown in Fig. 10.11(a). The optimised mesh consisted of about 760 quadratic thin shell elements (Mindlin shell element) and 55 contact elements with 2393 nodes in total. The mesh is much finer in the region close to the hole since large stress variations are expected; Fig. 10.11(b). The model, when run on a Sun Superstation 5 with 64 Mb main memory, takes just over 5 minutes CPU time.

In order to simulate the no contact region behind the bolt, a gap is left between 101° and −101°. For the rest of the interface special gap elements are employed which prevent penetration of the bolt elements into the plate elements, and allow the contact region between the bolt and plate to be modelled with friction. The load is applied at the centre of the bolt as a point load (linearly increasing from zero to 12 kN in 2 kN increments) along the x-axis. The radial degree of freedom of the boundary nodes in the bolt, between 90° and 270°, is fixed to avoid any rigid body movement.

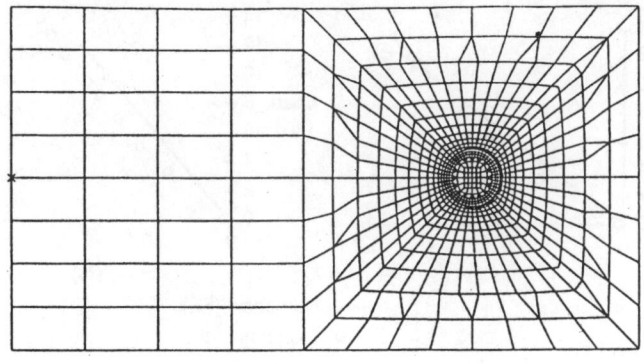

(a)

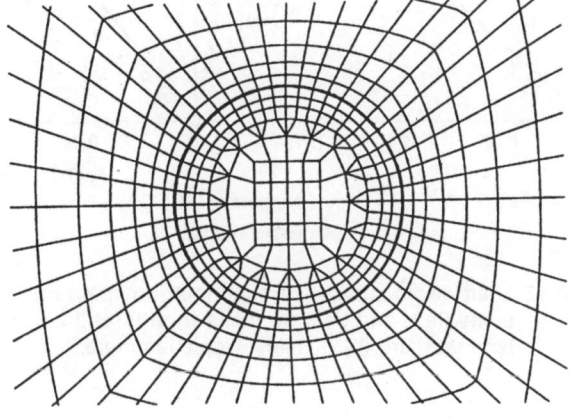

(b)

10.11 Plate with a pin-loaded hole: (a) overall mesh and (b) refined mesh around hole.

The FE load/strain curves corresponding to the net tension and bearing planes for the orthotropic and quasi-isotropic laminates are plotted in Fig. 10.12, and compared with experimental values obtained from strain gauges attached near the hole on a representative test specimen. Differences are due to the fact that the strain gauges measure an average value rather than a point value, and the FE analysis is 2D elastic, ignoring through-thickness effects.

For the (0/90°) orthotropic plate the maximum stress criterion predicts damage initiation at approximately 12 kN applied load, owing to high shear stress developed at $\theta = 45°$ from the direct contact point between the bolt and the plate. No damage is observed in the bearing plane; the compres-

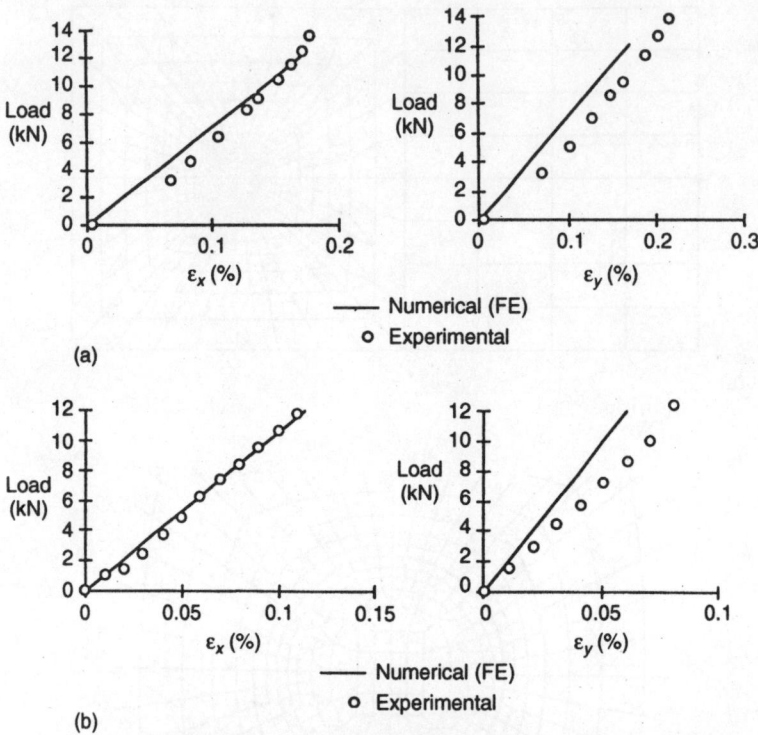

(a)

(b)

10.12 Load–strain curves corresponding to strain on the net-tension plane (left) and bearing plane (right): (a) the orthotropic laminate and (b) quasi-isotropic laminate.[32]

sive stress at the edge of the hole is below the critical compressive strength value of 520 MPa.

In the (0/90/±45°) laminate failure initiates first in the off-axis plies (±45° layers) at 45° from the direct contact point, owing to high shear, while the (0/90°) plies fail both in net-tension and bearing mode. A 12 kN applied load introduces stresses at the edges of the hole which exceed the ultimate tensile (650 MPa) and compressive strength (520 MPa) of the system, and therefore damage (net-tension and bearing) is expected to occur in these layers. The measured failure load for this lay-up is 16.5 kN, suggesting that the damage reduces the local stresses (load redistribution) and therefore delays final failure. The point stress failure criterion (PSFC) predicts net-tension failure in the quasi-isotropic laminate successfully. With the characteristic length d_0 from the hole edge taken equal to 1 mm, the predicted failure load is 16 kN compared with 16.5 kN for the measured value. For the orthotropic lay-up the theoretical failure load is 12 kN, which is very close to that observed (13kN). However, final failure occurred in the bearing plane.

With the PSFC, the linear elastic stress analysis is presumed to be valid outside some empirically determined softened zone adjacent to the hole. The basic drawback is that the so-called characteristic length varies considerably with bearing stress, and failure is predicted at some place other than where it is known to occur. The value chosen for d_0, although not determined analytically, correlates quite well with the value used for the strength of carbon fibre/epoxy laminates with an open hole.

10.3.5 3D FE analysis

As already mentioned, there are many factors that influence the performance of mechanical joints, and their complete representation, in any model, requires a 3D analysis. Necessarily, this will mean an FE model. In this section a particular approach is presented.[34]

10.3.5.1 Model formulation

Stacked, eight-node, reduced integration, brick elements were chosen as they are adequate to capture the 3D stresses around the hole boundary. The ABAQUS system[35] was used for the analysis, with I-DEAS[36] used as a preprocessor. The choice of element was also related to the contact algorithm used in ABAQUS. The mesh around the hole boundary is shown in Fig. 10.13; there is one element per ply thickness. Special care was taken to avoid element distortions due to aspect ratio and skew angular deformations. Nevertheless, the use of one element per ply will give high aspect ratios away from the fine mesh region, where accurate stresses are not so important. Because of symmetry, only one quarter of the laminate was modelled.

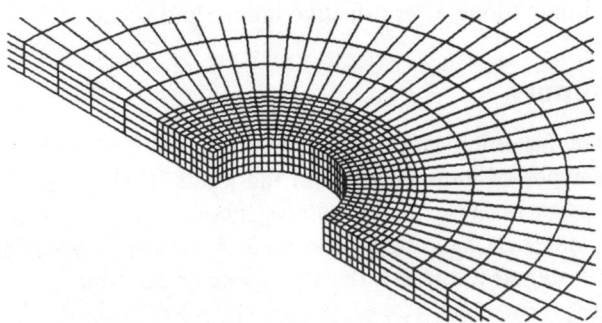

10.13 Typical 3D finite element mesh for a FRP laminate with a pin-loaded hole.[34]

The FE results were compared with experimental data obtained from single fastener specimens, with finger-tight washers, loaded in double shear. The washer was assumed to be rigid, and constraints imposed at the corresponding nodes in the model to prevent through-thickness expansion.

10.3.5.2 Representation of damage

Extensive experimental characterisation was carried out to identify the failure modes that occurred as the fastener load was increased from zero to final failure.[34] Matrix cracking in tension, compression and shear, fibre failure in tension and compression, and delamination were all seen. From this work it was concluded that a 3D failure criterion would be needed, together with a procedure for reducing the elastic properties in damaged elements. The latter was based on the approach of Tan and coworker,[37,38] and the criterion of Hashin[39] was used to indicate the onset and type of damage within a ply. The initiation of delamination was modelled, but not its growth.

Besides predicting the onset of non-critical damage, a procedure was also needed to indicate final failure. For the tension and shear out modes, total failure of the joint was taken to be when predicted fibre damage extended to the free edge of the laminate. For the bearing mode failure was when fibre damage reached the outer edge of the washer. These representations were in accord with experimental observations.

10.3.5.3 Comparison of predictions with experiment

To illustrate the success of the approach, the predicted and experimental load–displacement (from a linear voltage displacement transducer, LVDT) curves are shown in Fig. 10.14 for the bearing failure specmen. Comparisons for tension and shear out modes were equally good. There is significantly more extension from the experimental results simply because the tests were continued long after effective failure had occurred.

10.4   Summary

In this chapter, fastening methods commonly employed with composite materials, the types of joint failure and the kinds of problems that arise in the joint design because of the heterogeneous and anisotropic nature of composite materials have been discussed. Adhesive bonded joints and mechanically fastened joints are the two types of joint mainly used with these materials.

In the development of bonded joints for structures, a simple joint can be fabricated first and tested for its suitability in structures. The size of the joint

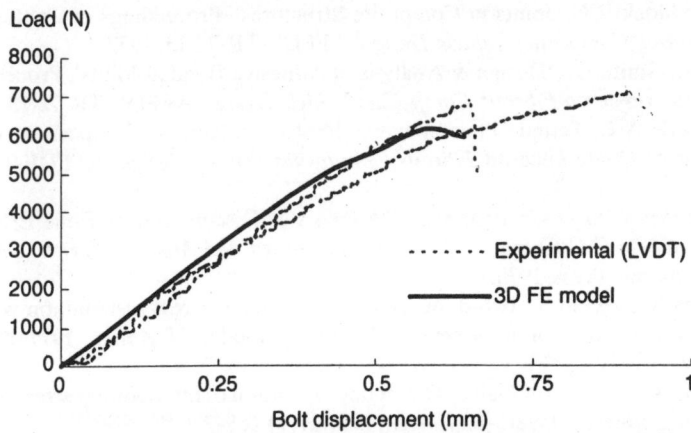

10.14 Experimental and predicted load–displacement relation for the bearing failure specimen.[34]

can be first estimated from a knowledge of the part sizes to be joined, the allotted space for the joint, and a general idea of how much overlap is required to carry the load. With such knowledge, preliminary joint designs can be made that can be refined using an iterative analysis procedure. The FE method is a powerful numerical tool that can be used to estimate direct and shear stress distributions required in the failure analysis.

In structures where parts are removed for inspection or maintenance bolted joints are required. A wide range of variables such as lay-up, fastener type (screw, rivet, bolt), friction effects, clearance and their influence on the failure mode of a bolted (or riveted) connection need to be carefully examined for a successful design. The stress analysis in such a joint must include their 3D nature, a fact that has limited the analytical treatment given to such connections. The prediction of failure loads is, for the moment at least, generally semi-empirical. However, it has been shown that the use of appropriate failure criteria that are more generally applicable, in combination with stiffness reduction laws, together with an easy-to-use 3D FE stress analysis can result in improved strength predictions.

10.5   References

1 Matthews F L & Rawlings R D, *Composite Materials: Engineering and Science*, Cambridge, Woodhead, 1999.
2 Goland M & Reissner E, 'The Stresses in Cemented Joints', *J Appl Mech*, 1944 A17–A27.
3 Berg K R, 'Problems in the Design of Joints and Attachments,' in Wendt F W, Liebowitz H and Perrone N, Eds., *Mechanics of Composite Materials*, New York, Pergamon, 1970.

4 Haddock R N, 'Joints in Composite Structures', Proceedings of Conference on *Fibrous Composites Vehicle Design*, AFFDL-TR-72-13, 1972.
5 Hart-Smith L J, 'Design & Analysis of Adhesive Bonded Joints', Proceedings of Conference on *Fibrous Composites Vehicle Design*, AFFDL-TR-72-13, 1972.
6 Fehrle A C, 'Fatigue Phenomena of Joints in Advanced Composites', Proceedings of Conference on *Fibrous Composites Vehicle Design*, AFFDL-TR-72-13, 1972.
7 Grimes G G & Greimann L F, 'Analysis of Discontinuities, Edge Effects, and Joints', in C C Chamis, Ed., *Structural Design and Analysis Part II*, New York, Academic Press, 1975.
8 Ishai O & Gali S, 'Two-dimensional interlaminar stress distribution within the adhesive layer of a symmetrical doubler model', *J Adhes*, 1977 **8**(4) 301–312.
9 Ishai O, Peretz D & Gali S, 'Direct determination of interlaminar stresses in polymeric adhesive layer', *Exp Mech*, 1977 **17**(7) 265–270.
10 Amijima S, Fujii T & Yoshida A, 'Two-Dimensional Stress Analysis on Adhesive Bonded Joints', Proceedings of Twentieth Japan Congress on *Materials Research*, The Society of Materials Science, Japan, 1977.
11 Amijima S, Fujii T, Yoshida A & Amino H, 'Dynamic Response of Adhesive Bonded Joints', Proceedings of Twentieth Japan Congress on *Materials Research*, The Society of Materials Science, Japan, 1977.
12 Hart-Smith L J, *Further developments in the design of adhesive-bonded structural joints*. Douglas Paper 6922, Douglas Aircraft Corp., Long Beach, CA, April, 1980.
13 Kairouz K C & Matthews F L, 'A finite element analysis of the effect of stacking sequence on the failure mode of bonded composite single lap joints', *Proc CADCOMP-2*, Brussels, Springer Verlag, 1990.
14 Matthews F L, Kilty P F & Godwin E W, 'A review of the strength of joints in fibre-reinforced plastics, Part 2: Adhesively bonded joints', *Composites*, 1982 **13**(1) 29–37.
15 Hitchings D, *FE77 Users' Manual*, Dept of Aeronautics, Imperial College, London, 1995.
16 Soutis C & Hu F-Z, 'A 3-D failure analysis of scarf patch repaired CFRP plates', 39th AIAA/ASME/ASCE/AHS/ASC Structures, *Structural Dynamics and Materials Conference*, Long Beach, CA, Paper No. AIAA-98-1943, **3** 1971–1977, 20–23 April, 1998.
17 Collings T A, 'The strength of bolted joints in multi-directional CFRP laminates', *Composites*, 1977 **8**(1) 43–54.
18 Collings T A, 'On the bearing strengths of CFRP laminates', *Composites*, 1982 **13**(3) 241–252.
19 Wong C M S & Matthews F L, 'A finite element analysis of single and two-hole bolted joints in fibre reinforced plastic', *J Compos Mater*, 1981 **15**(9) 481–491.
20 Matthews F L, Wong C M S & Chryssafitis C, 'Stress distribution around a single bolt in fibre-reinforced plastic', *Composites*, 1982 **13**(3) 316–22.
21 Kretsis G & Matthews F L, 'The strength of bolted joints in glass fibre/epoxy laminates', *Composites*, 1985 **16**(2) 92–102.
22 Agarwal B L, 'Static strength prediction of bolted joints in composite materials', *AIAA J*, 1980 **18**(11) 1371–1375.
23 Zhang K & Ueng C E S, 'Stresses around a pin-loaded hole in orthotropic plates', *J Compos Mater*, 1984 **18**(9) 119–143.

24 Smith P A, *Aspects of the Static and Fatigue Behaviour of Composite Laminates, Including Bolted Joints*, PhD Thesis, Cambridge University Engineering Department, 1985.

25 Eriksson I, 'Contact stresses in bolted joints of composite laminates', *Composite Structures*, 1986 6(1–3) 57–75.

26 Smith P A, Pascoe K J, Polak C & Stround D O, 'The behaviour of single-lap bolted joints in CFRP laminates', *Composite Structures*, 1986 6(1–3) 41–55.

27 Hyer M W, Klang E C & Cooper D E, 'Effects of pin elasticity, clearance and friction on the stresses in a pin-loaded orthotropic plate', *J Compos Mater*, 1987 21(3) 190–206.

28 Chen W H, Lee S S & Yeh J T, 'Three-dimensional contact stress analysis of a composite plate with a bolted joint', *Composite Structures*, 1995 30(3) 287–297.

29 Hart-Smith L J, 'Mechanically-fastened joints for advanced composites-phenomenological considerations and simple analyses', Proc. 4th Conf. on *Fibrous Composites in Structural Design*, San Diego, USA, 1978.

30 Camanho P P & Matthews F L, 'Stress analysis and strength prediction of mechanically fastened joints in FRP: A review', *Composites Part A*, 1997 28A(6) 529–547.

31 Lekhnitskii S G, *Theory of Elasticity of an Anisotropic Elastic Body*, San Fransisco, Holden-Day Inc., 1963.

32 Andreasson N, Mackinlay P & Soutis C, 'Tensile strength of bolted joints in woven CFRP laminates', *The Royal Aeronautical J*, 1998 102(1018) 445–450.

33 Whitney J M & Nuismer R J, 'Stress fracture criteria for laminated composites containing stress concentrations', *J Compos Mater*, 1974 8(7) 253–265.

34 Camanho P P, *Application of Numerical Methods to the Strength Prediction of Mechanically Fastened Joints in Composite Laminates*, PhD Thesis, Centre for Comp Matls, Imperial College, London University, 1999.

35 *ABAQUS 5.6 Users' Manual*, Hibbitt, Karlsson & Sorensen Inc, Pawtucket, USA.

36 *I-DEAS Master Series Release 2 Users' Manual*, Structural Dynamics Research Corporation, Milford, Ohio.

37 Nuismer R J & Tan S C, 'Constitutive relations of a cracked composite lamina', *J Compos Mater*, 1988 22(4) 306–321.

38 Tan S C & Nuismer R J, 'A theory for progressive matrix cracking in composite laminates' *J Compos Mater*, 1989 23(10) 1029–1047.

39 Hashin Z, 'Failure criteria for unidirectional fiber composites', *J Appl Mech*, 1980 47(2) 329–334.

11
Fatigue

11.1 Introduction

Fatigue in metals occurs by the initiation of a single crack and its intermittent propagation until catastrophic failure occurs with little warning and no sign of gross distortion, even in highly ductile metals, except at the final tensile region of fracture. In contrast to homogeneous materials, composites accumulate damage in a general rather than a localised fashion, and fracture does not always occur by propagation of a single macroscopic crack. The microstructural mechanisms of damage accumulation, including fibre/matrix debonding, matrix cracking, delamination and fibre fracture, occur sometimes independently and sometimes interactively, and the predominance of one or other of them may be strongly affected by both materials' variables and testing conditions.[1,2] Fatigue damage in composites leads to permanent degradation in mechanical properties, notably laminate stiffness and residual strength, often at an early stage in the fatigue life.

In this chapter, we describe the fatigue damage accumulation in unidirectional and multidirectional fibre-reinforced polymer laminates, and discuss some of the methods which have been used to model composite fatigue behaviour. A more elaborate discussion on the subject can be found in references 1–7.

11.2 Damage modes in composite laminates

In multidirectional laminates, the 0° plies carry most of the load and provide most of the stiffness while the 90° and +45° plies give transverse and shear strength and stiffness, respectively. Under an applied load (mechanical, thermal, static or cyclic) a complicated state of damage develops in the off-axis plies, causing load redistributions which lead to eventual fracture of the load-bearing plies.

Damage in (0/90°) cross-ply and (0/90/±45°) quasi-isotropic laminates is of three main types: matrix cracking parallel to the fibres in the longitudinal and off-axis plies, delamination between plies and fibre fracture with associated debonding. The general sequence of damage is similar under tensile static and fatigue loading. Fibre/matrix debonds initiate around fibres lying at an angle to the loading direction, initially in the 90° plies, extend to form microcracks which, in turn, form matrix cracks across the thickness and the width of the plies. The number of cracks in all of the off-axis plies increases with increasing static stress or number of fatigue cycles, reaching a maximum density which remains stable until fracture. This has been termed the 'characteristic damage state' (CDS) and depends on ply thickness and orientation and on laminate stacking sequence.[8]

Matrix cracking parallel to the fibres in the 0° plies occurs at strains close to failure under static loading or may develop throughout fatigue cycling (owing to the mismatch between Poisson's ratios of adjacent plies).[9,10]

Delaminations initiate between plies of different orientation and grow inwards from the laminate edges across the width of a test coupon. The tendency to delaminate results from out-of-plane interlaminar shear and normal stresses which exist at free edges and depends on the stacking sequence of the off-axis plies within the laminate,[9] as described in Chapter 7.

Final failure of the laminate is due to tensile fracture of the 0° fibres. The variation in fibre strength is related to the statistical distribution of flaws along the fibre length. Progressive fibre failure and debonding of the broken fibres (due to the high shear stresses at the fibre/matrix interface close to the break) causes a redistribution of the applied load, increasing the probability of fracture in neighbouring fibres and reaching a critical number of fibre failures to cause fracture. A higher concentration of fibre breaks has been observed at the intersection of transverse ply cracks with the longitudinal plies under cyclic loading.[11]

In compression, although the fibres remain the principal load-bearing elements, they must be supported from becoming locally unstable and undergoing a microbuckling type of failure.[4,12] This is the task of the matrix and the fibre/matrix interface, the integrity of both being of greater importance in compressive loading than in tensile loading. Matrix and interfacial damage develop for much the same reasons as for tensile loading, but because of the greater demand on the matrix and the interface in compressive loading, compressive fatigue loading generally has a greater effect on the residual strength of composite materials than does tensile loading. In addition, local resin and interfacial damage leads to fibre instability in compressive loading which is more severe than the fibre isolation mode which occurs in tensile loading. Generally, fewer studies are available on the compressive fatigue of composites, mainly because compression testing of these materials presents many problems, not the least of which is the need to support the specimen from undergoing global macrobuckling, com-

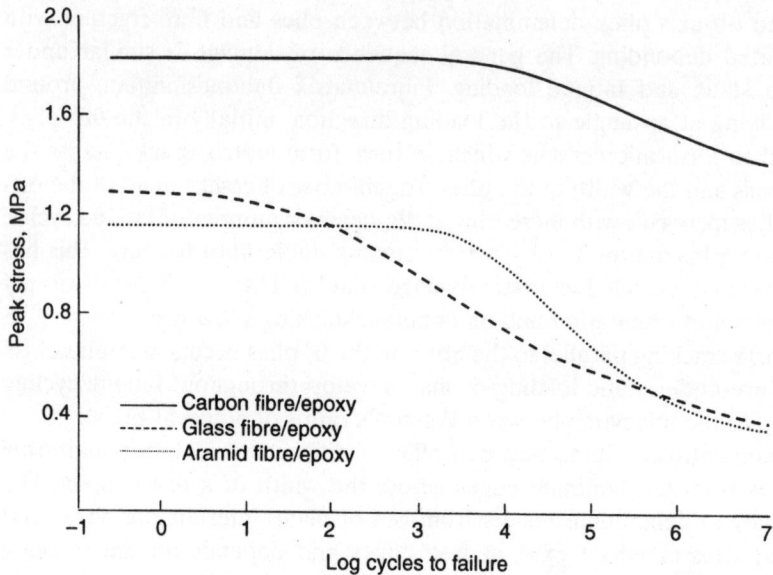

11.1 S-N fatigue data for unidirectional composite materials.[4]

bined with limitations imposed on specimen geometry by the anisotropic nature of the materials.[4]

11.3   Fatigue of unidirectional laminates

11.3.1 Fatigue lifetime

Typical plots of peak tensile stress versus log cycles to failure, the traditional stress-cycles (S–N) presentation of data, are shown in Fig. 11.1.[4] Data for three unidirectional materials are presented: carbon fibre, glass fibre and aramid fibre-reinforced epoxy resin. The slope and shape of the S–N curve is a measure of the fatigue resistance of the material.

Since, for unidirectional composite materials under tensile loading, the fibres carry virtually all the load, the tensile fatigue behaviour might be expected to depend solely on the fibres, and since the fibres themselves are not usually particularly sensitive to fatigue loading, good fatigue behaviour should result. However, experimental evidence has shown that the slopes of the curves are determined principally by the strain in the matrix[13,14] when the fatigue limit of the matrix is less than that of the fibres, which is nearly always the case. Consequently, plots of mean strain rather than stress vs log cycles to failure are frequently more meaningful for composite materials. The resulting ε–N diagram, Fig. 11.2, would show regions of dominance of

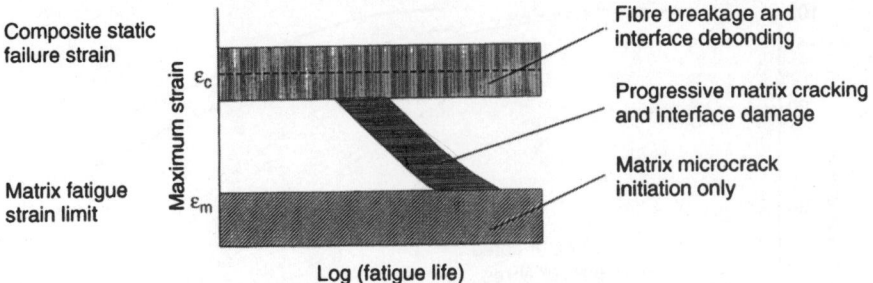

11.2 A generalised strain–fatigue life diagram for unidirectional
composites loaded parallel to the fibres showing dominant
regions of tensile fatigue damage.[14]

individual damage mechanisms in the fibre, matrix and interface from which
the contribution of each to the overall fatigue response can be evaluated.[14]

In unidirectional fibre/epoxy laminates, fatigue failures tend to occur only
within a narrow band of stress close to the static strength of the laminate,
and the number of cycles to failure can vary over several decades of cycles
inside this band. Load is carried mainly by the stiff fibres and there is little
permanent damage until fibre fracture is initiated. Fatigue damage may
accumulate as a result of stress redistribution from broken fibres but this is
only significant at stresses close to the static ultimate strength.

Comparison of the S–N curves shown in Fig. 11.1 illustrates that fatigue
performance is ranked in the order of modulus and strain to failure of
the fibres. The use of very stiff fibres, such as carbon fibres (typically with
modulus of 220–700 GPa and failure strains of 0.6–2%), limits the strain in
the composite and so prevents large elastic and visco-elastic deformations
in the matrix which lead to initiation of damage. The lower modulus of glass
fibres (70–80 GPa) permits composite strains large enough to cause early
matrix damage and hence precipitate fatigue failure. Consequently, the S–N
curve falls more steeply than for CFRP. The fatigue performance of Kevlar
fibre composites is intermediate to CFRP and GFRP. However, the fatigue
damage mechanism is complicated in this material since aramid fibres are
themselves fatigue sensitive and can defibrillate during fatigue loading.
The low-cycle behaviour compares well with GFRP but deteriorates at
intermediate to long lifetimes.

11.3.2 Fatigue damage accumulation

Some of the weakest fibres fail in the first few cycles since the cyclic stress
falls within the statistical fibre strength distribution determined by flaws.[15]
This gives rise to locally enhanced stresses in the matrix and at the

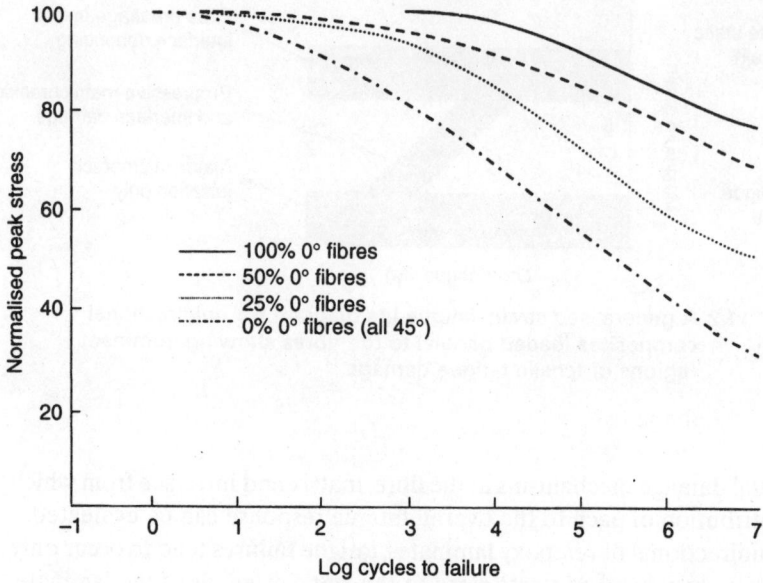

11.3 Normalised S–N curves for (0/±45°) CFRP laminates with varying percentage of 0° fibres.[4]

fibre/matrix interface, which lead to the development of fatigue damage with increasing number of cycles. Damage may also develop at local microdefects, such as misaligned fibres, resin-rich regions or voids.[4] Resin cracks frequently develop between the fibres, isolating them from adjacent material and rendering them ineffective load carriers, causing fibres to become locally overloaded and further fibre failures to occur. Close to failure, the matrix may show extensive longitudinal splitting parallel to fibres caused by resin and interfacial damage, leading to the brush-like failure characteristic of most unidirectional materials.[4] The rate of this degradation process in the matrix and at the interface is a function of the bulk strain in the resin as well as the nature of the matrix.

11.4   Fatigue of multidirectional laminates

11.4.1 Fatigue lifetime

On increasing the percentage of non-axial plies in a laminate, the static tensile strength and stiffness are reduced since fewer fibres are available to support the mean applied loads. The slope of the tensile S–N curve increases in relation to the static strength, Fig. 11.3, as the layers with off-axis fibres, whose mechanical properties are resin-dependent, are more easily damaged in fatigue.[4]

Both CFRP and GFRP multidirectional laminates undergo progressive accumulation of fatigue damage at rates that are reflected in the slope of the S–N curves, as explained in Section 11.2. The damage will lead to a general loss of integrity with potential for environmental attack, and certainly to a reduction in compressive strength.[15,16] Ultimate tensile fatigue failure of composite laminates is usually determined by the 0° layers; thus the tensile S–N curves for multidirectional laminated composite materials are still relatively shallow, although steeper than for 100% unidirectional material. In general, the poorer fatigue performance observed in unidirectional materials through the use of tough matrices is less marked in multidirectional laminates.

11.4.2 Progressive damage in multiply laminates

The general sequence of damage events in a multidirectional laminate under fatigue loading is similar to that observed under static loading, as explained above. At high fatigue stresses, cracks can initiate on the first loading cycle and then accumulate with increasing number of cycles. However, cracks can develop even when the maximum cyclic stress is well below the static cracking threshold, although only after an 'incubation period' of hundreds of thousands of cycles (depending on peak stress); Fig. 11.4.[17] Cracks generally initiate at the free edge of the specimen but can also initiate away from the edge. Transverse ply cracks are observed to grow stably across the width of the ply under fatigue loading at a rate which depends on the cyclic stress level and on interaction with neighbouring cracks.

The early initiation of matrix cracking in fatigue relative to static loading consequently leads to a decrease in the threshold for the onset of other types of damage. Delaminations can propagate over many thousands of cycles, resulting in separation of the laminate into discrete laminae (which will continue to support tensile load via the 0° plies).

Analysis of the stresses at the intersection of matrix cracks in adjacent plies of different orientation[18] shows that there is a highly localised region of increased interlaminar normal and shear stress around the point of intersection, which could be an initiation point for failure of the laminate. The final stage of damage development is dominated by fibre failures that result from these locally enhanced stresses.[19,20] Ultimately fracture occurs when the locally failed regions have sufficiently weakened the laminate to cause failure at the maximum applied load.

The progression of damage in woven fabric composites in the initial stage of tension–tension fatigue shows basically the same features that are

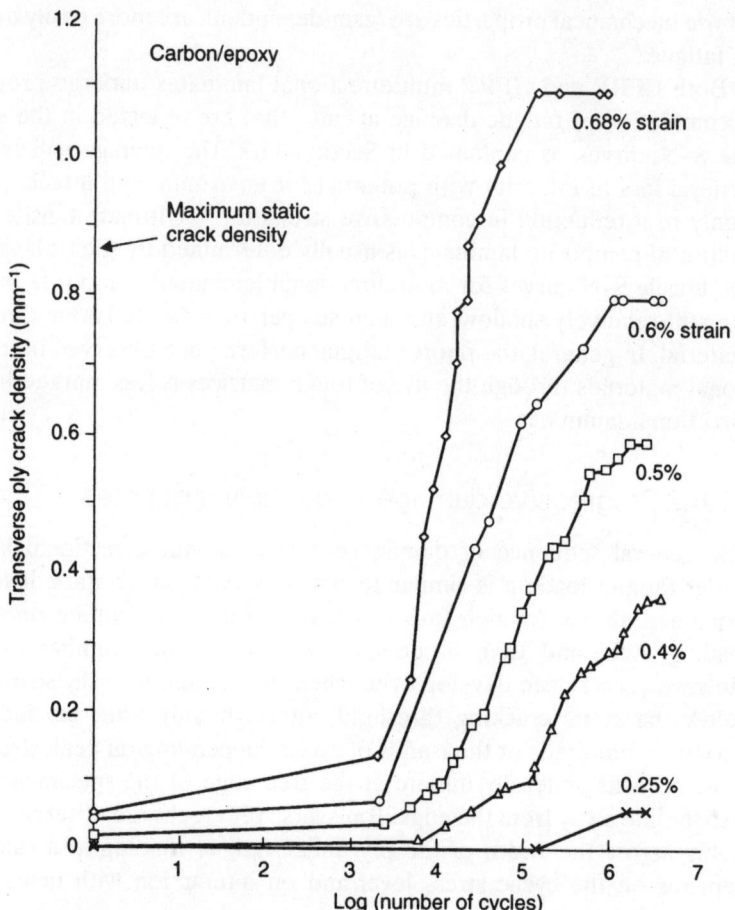

11.4 Variation of transverse ply crack density with cycles for CFRP (0/90°₃)ₛ specimens, fatigued at maximum stresses between 105 and 285 MPa (static cracking threshold, 0.45%).[17]

observed in cross-ply laminates,[21] i.e. cracks initiate in the weft (transverse) direction along the fibres. Further on in the lifetime, longitudinal cracks appear between warp fibres in the undulation region where the fibres cross over. Crack densities increase with cycling and a uniform pattern of orthogonal cracks develops. After further cycling, delaminations occur confined to the undulation regions and, thereby, attain a uniform distribution in the interlaminar planes. Towards the end of the fatigue life fibre bundles fail in the undulation regions and lead to general weakening and subsequent failure. The S–N curve falls more steeply than for equivalent non-woven material.[4] Thus, although woven composites have many processing advantages, their mechanical properties, particularly in fatigue loading, are generally poorer than unidirectional tape materials.

11.5   Fatigue damage effects on mechanical properties

11.5.1 Stiffness

The progressive damage sustained by a laminate during fatigue will affect the macroscopic mechanical properties of the material to an extent which depends on the specimen geometry, laminate stacking sequence and mode of testing. For example, in unidirectional, high-modulus carbon fibre/epoxy laminates, subjected to tension–tension fatigue, no significant change in composite stiffness is expected prior to failure. By contrast, in cross-ply laminates transverse ply cracking occurs early in the life of the specimen, causing a stiffness reduction that will increase with number of cycles.

Stiffness reduction in composite laminates has been studied extensively and very good correlations have been obtained with the level of matrix cracking.[22] The curve of normalised modulus versus cycles consists of three distinct regions;[18] Fig. 11.5. The initial rapid stiffness reduction of Stage I is

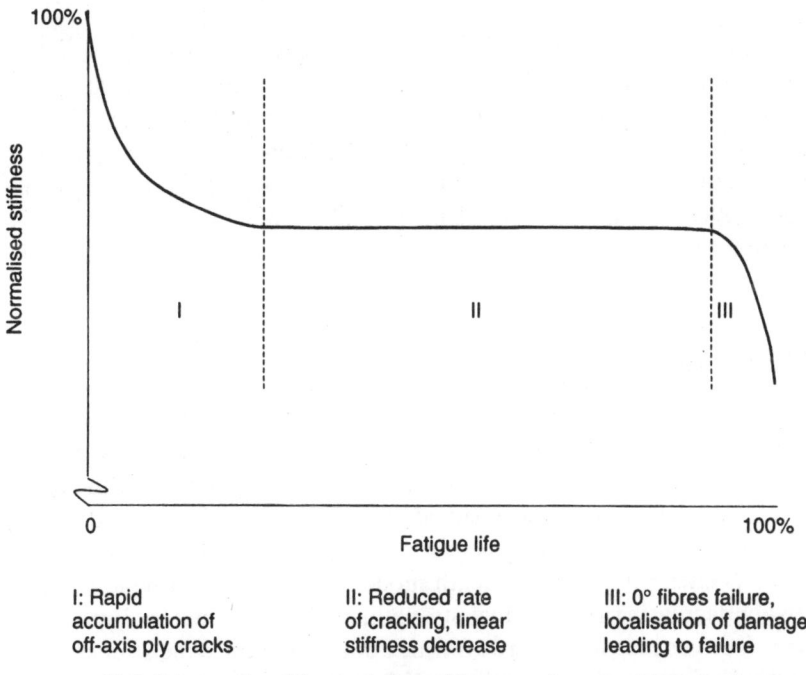

I: Rapid accumulation of off-axis ply cracks

II: Reduced rate of cracking, linear stiffness decrease

III: 0° fibres failure, localisation of damage leading to failure

11.5 Schematic stiffness–fatigue life curve for a multiply laminate showing three characteristic stages of stiffness reduction.[18]

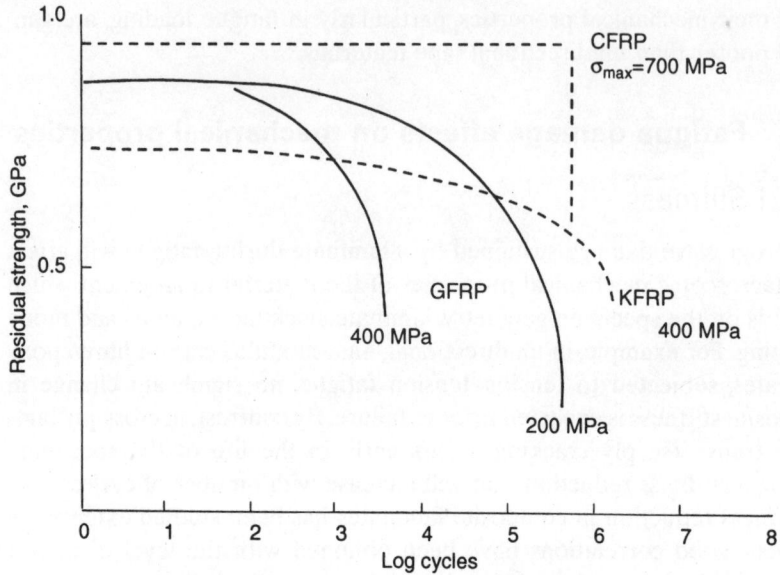

11.6 Residual strength of three kinds of 0/90° laminates following fatigue cycling at the stress level indicated: $R = 0.1$.[2]

dominated by the formation of off-axis ply cracks, although some edge delamination (in a 0/90/45° laminates) and 0° fibre fracture (in 0/90/0°) may also occur. Throughout the approximately linear region of Stage II, the matrix crack density is still increasing but much more slowly, and further fibre fracture is accompanied by 0° ply splitting (0/90/0°) and edge delaminations (0/90/45°). In Stage III, the final rapid stiffness reduction is associated with localisation of the damage processes to form a macro-crack which precipitates ultimate failure.

11.5.2 Residual strength

The effect of fatigue loading on laminate residual strength can be described by the concepts of 'wear-out' and 'sudden death'. The wear-out model assumes that the residual strength of the laminate falls gradually with cycles until failure occurs when the residual strength equals the applied cyclic stress; Fig. 11.6. This behaviour is observed in cross-ply GFRP and Kevlar fibre composites.[23] The sudden-death model proposes that instead of progressively decreasing the residual strength, some other material property changes (such as matrix properties) which does not reduce the strength significantly until very close to failure. This type of behaviour is generally shown by CFRP laminates; Fig. 11.6.

11.6   Fatigue damage modelling

Although the details of the damage mechanisms in the various composite systems are quite varied, it is possible to identify certain general features in the damage development. As an overall generalisation damage development may be considered to take place in two stages: the distributed damage stage and the localised damage stage.[24] Furthermore, the distributed damage stage is initially non-interactive, i.e. the individual damage entities initiate and grow uninfluenced by the others. The degree of interaction increases, first gradually, and eventually becomes severe enough to cause localised damage development in some preferred zones. Further development of damage involves instability and loss of integrity, and is statistically governed. The present state of damage modelling, using a micromechanics[25-27] or continuum damage approach,[28,29] has come as far as to model the distributed damage stage. Localised damage development is not yet sufficiently understood to attempt realistic modelling, including, for example, multiple delaminations.

Earlier studies on damage accumulation during cyclic loading attempted to obtain useful working relationships of the Miner's rule type, that might be used in design. Howe and Owen[30] who studied CSM/polyester composites suggested that although debonding did not itself cause reductions in strength it did serve to initiate resin cracks which weakened the material. For resin cracking during fatigue, the authors proposed a non-linear damage law, independent of stress level, which gives the damage, Δ, as

$$\Delta = \Sigma \left[A\left(\frac{n}{N}\right) + B\left(\frac{n}{N}\right)^2 \right]$$ [11.1]

where n is the number of cycles sustained by the composite at a stress level which would normally cause failure after N cycles, and A and B are constants. Δ is equal to unity at failure. Fong[31] proposed an alternative single-parameter stress-independent damage model in which the damage factor Δ was given by

$$\Delta = \frac{(e^{kx} - 1)}{(e^k - 1)}$$ [11.2]

for $k \approx 0$ and $0 < x < 1$, where k and x are constants determined from experimental data. This model implies that the rate of accumulation of damage (matrix cracking), $d\Delta/dt$, depends linearly on t as a first approximation, t being the non-logarithmic cycle ratio n/N. For different values of k this function generates a family of exponential curves.

A similar approach to quantify and predict the effect of damage on fatigue life was used by Beaumont and coworkers,[32] where all the individual damage mechanisms, e.g. splitting, delamination and matrix cracking, are assumed to contribute to a global damage parameter, Δ. This parameter can be related to a measurable change in laminate property such as stiffness to find the rate of change of Δ with cycles and then integrated to give Δ after N cycles at a given load. A suitable failure criterion is then applied to determine the critical value of Δ.

Also, the rate of growth of individual types of damage (e.g. transverse ply cracking) have been considered. A fracture mechanics parameter, the strain energy release rate G, has been used to relate fatigue crack growth to the change in laminate compliance[33] using the relationship between G and the change in laminate compliance with transverse ply crack length. Wang et al.[34] assumed a simple power law dependence of the rate of growth of full ply-width flaws across the ply thickness on the strain energy release rate (calculated by FE analysis). The energy release rate depends on crack spacing and is reduced by crack interactions, which is relevant to experimental observations of the effect of the proximity of neighbouring cracks on growth rate as the crack density increases.[35] The strain energy release rate has also been used to model fatigue delamination growth in angle-ply laminates.[36]

A different fracture mechanics approach to modelling matrix crack growth is taken by Bader et al.[17] in which an approximate expression is derived for the stress intensity factor at the tip of a full ply-thickness transverse crack growing across the ply width. The stress intensity factor also depends on crack spacing and decreases as the spacing decreases. A Paris Law relationship gave good correlation between total crack growth rate (inferred from the measured rate of stiffness reduction) and the maximum stress intensity factor.

In some recent work, Behesty et al.[37] have developed a constant-life model which is able to describe the stress/life/R-ratio failure surface of a given laminate in terms of the monotonic tensile and compressive strengths of the material and three empirical parameters, referred to as f, u and v. It was found that for a CFRP laminate the values of u and v are close to 2 and do not vary greatly with life or material characteristics. The parameter, f, on the other hand, is a function of the ratio of the compressive and tensile strengths, and varies with life. The model appears to be valid for glass fibre-reinforced plastics as well as for the carbon fibre composites for which it was originally developed.

11.7 A finite element model for fatigue life prediction of CFRP components

11.7.1 Introduction

Very little FE modelling of the fatigue of composite structures appears to have been undertaken. In this section a particular case of a web-notched

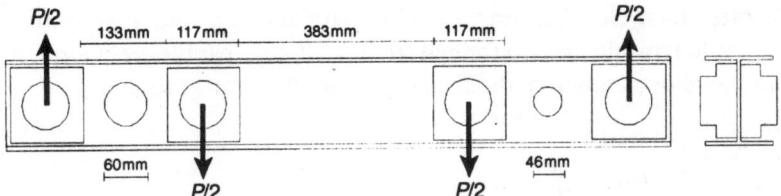

11.7 A web-notched I-beam subjected to a four-point bending load.[38]

I-beam subjected to the four-point bending load illustrated in Fig. 11.7 is described.[38] The basis of the method is first to use a fracture mechanics approach to obtain experimentally the relationship between the maximum strain energy release rate experienced during a fatigue cycle and the rate of growth of the damage. For the beam, the latter was mainly matrix cracking, as determined by observations during a series of fatigue tests. The component is then analysed using FE analysis, the local stresses and strains around the notch determined, the energy release rate calculated and then combined with the fracture mechanics data to give the lifetime, which is compared with experimental results.

11.7.2 Outline of predictive method

The model is based on a combination of FE stress analysis and fracture mechanics. The FE analysis (using the ABAQUS system, with 2D shell elements) is used to get a detailed stress distribution in the structure when subjected to a static load equal to the maximum load applied in the fatigue cycle. Interrogation of the output allows identification of strains above a designated threshold (for matrix cracking), and the elements and plies within the element in which these strains occur.

The elastic properties in the damaged plies are then reduced and the analysis repeated. Further damage is identified and the process continued until final failure is indicated. In the experimental work fibre fracture was identified as being the ultimate cause of failure. Hence, in the model, fibre strains are monitored and when the ultimate fibre failure strain is reached, at any location, the analysis is halted.

At each iteration of the analysis the strain energy release rate, G, in the structure is calculated. So, at the conclusion of the analysis a relationship between G and damage area, A, is available. This information is then combined with fracture mechanics in the form of the Paris Law, and from this the lifetime is found.

An initial model was based on the simple Paris Law: $dA/dN = DG_{max}^n$ linking damage area, A, number cycles, N, and maximum strain energy release rate in a cycle, G_{max}, via 'materials' parameters D and n. This information would typically be determined from the double cantilever beam test.

From the above equation, the lifetime is found by integration as

$$N_f = \int_0^{A_f} \frac{dA}{D[G(A)]^n}$$

where the final damage area A_f corresponds to fibre fracture.

The lifetime predictions from the initial model proved to be extremely sensitive to values of n and D and to overcome this a modified form of the Paris Law was used:

$$\frac{dA}{dN} = DG_{max}^n \frac{\left[1 - \left(\frac{G_{th}}{G_{max}}\right)^{n_1}\right]}{\left[1 - \left(\frac{G_{max}}{G_c}\right)^{n_2}\right]}$$

where G_{th} is a threshold value of G_{max}, below which no cracking occurs, and G_c is the critical value above which instantaneous failure occurs. The terms n_1 and n_2 are additional 'materials' parameters. The integration to get lifetime was performed numerically. Data for stiffness reduction due to matrix cracking were obtained from published work on a similar material to that from which the beams were fabricated (CFRP: XAS/914 from Hexcel Composites).[39]

11.7.3 Results from model

A key issue proved to be the availability, and quality, of fracture mechanics data, essentially the Paris Law constants (n and D (and n_1 and n_2 if appropriate)). Owing to scatter in the results of DCB, and similar tests, a spread of n and D is always apparent. The values chosen, by curve fitting the experimental points, have a major influence on the predictions.

Another factor that influences the outcome of the predictions is the values of elastic constants that are input to the FE model. The most critical parameter seems to be the modulus (E_{22}) at right angles to the fibres in the unidirectional ply, although Poisson's ratio and shear modulus also play a part.

The best estimate, from the modified Paris Law, was 3.3×10^6 cycles to failure, which compares with the experimental value for the lifetime, N_f, of 4.78×10^6 cycles.

The predicted growth of damage area compared well with experiment. A typical FE mesh is shown in Fig. 11.8 and a comparison of predicted and measured damage growth is seen in Fig. 11.9.

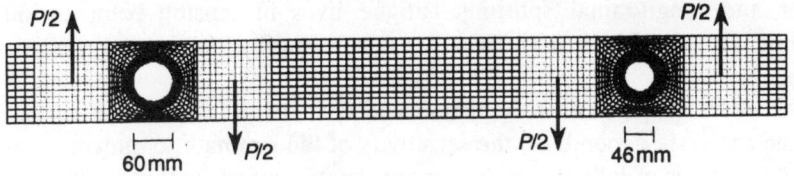

11.8 FE mesh of notched I-beam.[38]

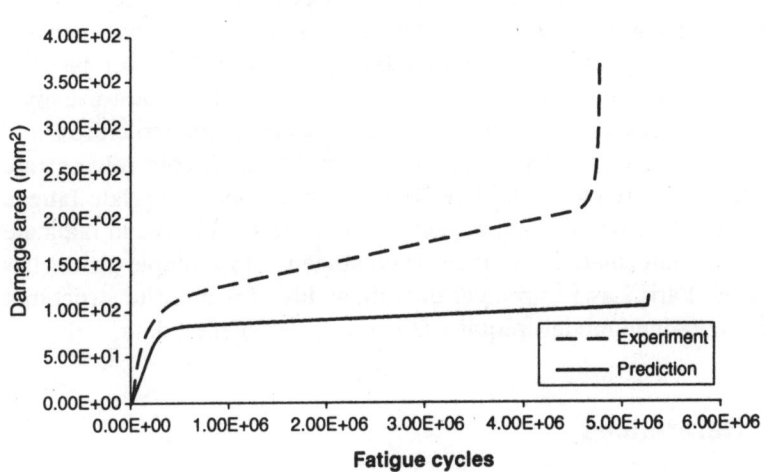

11.9 Comparison of experimental and predicted damage growth in
I-beam.[38]

11.8   Summary

In contrast to metals, composites accumulate damage in a general rather
than a localised fashion, and fracture does not usually occur by propagation of a single macroscopic crack. The microstructural mechanisms of
damage accumulation, including fibre/matrix debonding, matrix cracking,
delamination and fibre fracture, occur sometimes independently and sometimes interactively, and the predominance of one or other of them may
be strongly affected by both materials' variables and testing conditions.
Fatigue damage leads to a permanent degradation in laminate stiffness and
residual strength.

The fatigue resistance of composites is reduced when the applied load is
other than tensile and in the fibre direction. Compression, flexure and
torsion subject the relatively weak matrix and fibre/matrix interface to
shear stresses, reducing the contribution of the fibres to the fatigue resistance and allowing compressive fibre buckling initiated by local matrix

shear and longitudinal splitting. Fatigue lives in tension–compression loading are usually shorter than for zero-compression or zero-tension fatigue.

The effect of exposure to moisture and changes in temperature on the fatigue response depends on the sensitivity of the laminate to matrix properties, since it is usually the matrix or fibre/matrix interface that is affected. Thus carbon fibre laminates, having a strong fibre/matrix interface, show little sensitivity to moisture content or a rise in temperature.

It should be noted that the test method, specimen design, edge effects and stress concentrations such as notches, holes, fasteners, impact damage and other imperfections may affect the fatigue strength. However, because of relatively low static design allowables, fatigue strength is not usually a serious design criterion for current carbon fibre composite structures.

Finally, a fracture mechanics parameter, the strain energy release rate (calculated by finite element analysis) is commonly used to relate fatigue crack growth (matrix cracking or delamination) to the change in laminate stiffness. This information can then be combined with a simple power law (usually the Paris Law) to predict the fatigue life of composite structures. All existing fatigue models require extensive experimental data.

11.9 References

1 Harris B, 'Fatigue and accumulation of damage in reinforced plastics', *Composites*, 1977 **8**(4) 214–220.
2 Harris B, *Engineering Composite Materials*, London, Institute of Metals, 1986.
3 Konur O & Matthews F L, 'Effect properties of the constituents on the fatigue performance of composites: a review', *Composites*, 1987 **20**(4) 317–328.
4 Curtis P T, 'The fatigue behaviour of fibrous composite materials', *J Strain Analysis*, 1989 **24**(4) 47–56.
5 Talreja R, *Fatigue of Composite Materials*, Ed L L Reifsinider, New York, Elsevier Science Publishers, 1991.
6 O'Brien T K (Ed), *Composite Materials: Fatigue & Fracture*, 3rd Volume, ASTM STP-1110, American Society for Testing and Materials, Philadelphia, PA, 1991.
7 Carlson R L & Kardomateas G A, *An Introduction to Fatigue in Metals and Composites*, London, Chapman & Hall, 1996.
8 Reifsnider K L & Jamison R D, 'Fracture of fatigue-loaded composite laminates', *Int J Fatigue*, 1982 **4**(4) 187–197.
9 Pagano N J & Bryon Pipes R, 'Influence of stacking sequence on laminate strength', *J Compos Mater*, 1971 **5**(1) 50–57.
10 Bailey J E, Curtis P T & Parvizi A, 'On the transverse cracking and longitudinal splitting behaviour of glass and carbon fibre-reinforced epoxy cross-ply laminates and the effect of poisson and thermally generated strain', *Proc Royal Society London*, 1979 **A366**(1727) 599–623.
11 Jamison R D, 'The role of microdamage in tensile failure of graphite/epoxy laminates', *Composites Sci Tech*, 1985 **24**(2) 83–99.

12 Purslow D, 'Some fundamental aspects of composites fractography', *Composites*, 1981 **12**(4) 241–247.

13 Curtis P T, 'An investigation of the tensile fatigue behaviour of improved carbon fibre composites materials', *Proc ICCM-6*, Elsevier Applied Science London, 1987.

14 Talreja R, 'Fatigue of composite materials: damage mechanisms and fatigue life diagrams', *Proc Royal Society London*, 1981 **A387**(1775) 461–475.

15 Jamison R D, Schulte K, Reifsnider K L & Stinchcomb W W, 'Characterisation and analysis of damage mechanisms in fatigue of graphite/epoxy laminates', ASTM STP-836, American Society for Testing and Materials, Philadelphia, PA, 1983.

16 Reifsnider K L & Jamison R D, 'Fracture of fatigue loaded composite laminates', *Int J Fatigue*, 1982 **4**(4) 187–198.

17 Bader M G & Boniface L, 'Damage development during quasi-static and cyclic loading in GRP and CFRP laminates containing 90° plies', *Proc Conf ICCM-V*, The Metallurgical Society Inc, Warrendale, PA, 1985.

18 Highsmith A L & Reifsnider K L, 'Internal load distribution effects during fatigue loading of composite laminates', *Debonding of Materials* ASTM STP-907, Philadelphia, PA, American Society for Testing and Materials, 1986.

19 Schulte K, 'Development of microdamage in composite laminates during fatigue loading', *Proc Int Conf Testing, Evaluation & Quality Control of Composites*, Butterworths, 1983.

20 Jamison R D, 'On the interrelationship between fibre fracture and ply cracking in graphite/epoxy laminates', *Debonding of Materials* ASTM STP-907, Philadelphia, PA, American Society for Testing and Materials, 1986.

21 Curtis P T & Bishop S M, 'An assessment of the potential of woven carbon fibre reinforced plastics for high performance use', *Composites*, 1984 **15**(4) 259–265.

22 Highsmith A L & Reifsnider K L, 'Stiffness reduction mechanisms in composite laminates', ASTM STP-775, American Society for Testing and Materials, Philadelphia, PA, 1982.

23 Adam T, Dickson R F, Jones C J, Reiter H & Harris B, 'A power law fatigue damage model for fibre-reinforced plastic laminates', *Proc Inst Mech Engs*, 1986 **200**(3) 155–166.

24 Talreja R, 'Damage development in composites: mechanisms and modeling', *J Strain Analysis*, 1984 **24**(4) 27–34.

25 Laws N, Dvorak G J & Hejazi M, 'Stiffness changes in unidirectional composites caused by crack systems', *Mech Materials*, 1983 **2**, 123–137.

26 Hashin Z, 'Analysis of cracked laminates: a variational approach', *Mech Materials*, 1985 **4**, 122–136.

27 Aboudi J, 'Stiffness reduction in cracked solids', *Eng Fracture Mech*, 1987 **26**(5) 637–650.

28 Talreja R, 'A continuum mechanics characterisation of damage in composite materials, *Proc Roy Soc*, London, 1985 **A399**(1817) 195–216.

29 Allen D H, Groves S E & Harris C E, 'A cumulative damage model for continuous fibre composite laminates with matrix cracking and interply delaminations', ASTM STP-972, Philadelphia, PA, American Society for Testing and Materials, 1988.

30 Howe R J & Owen M J, *Proc Eighth Int Reinforced Plastics Congress*, London, British Plastics Federation, 1972.

31 Fong J T, *Damage in Composite Materials*, ASTM STP-775, Ed K L Reifsnider, Philadelphia, PA, American Society for Testing and Materials, 1982.

32 Poursartip A, Ashby M F & Beaumont P W R, 'The fatigue damage mechanisms of a carbon fibre composite laminate: Development of the model', *Compos Sci Tech*, 1986 **25**(3) 193–218.

33 Beaumont P W R, 'The failure of fibre composites: an overview', *J Strain Analysis*, 1989 **24**(4) 1–17.

34 Wang A S D, Chou P C & Lei S C, 'A stochastic model for the growth of matrix cracks in composite laminates', *J Compos Mater*, 1984 **18**(3) 239–254.

35 Boniface L & Ogin S L, 'Application of the Paris equation to the fatigue growth of transverse ply cracks', *J Compos Mater*, 1989 **23**(7) 735–754.

36 O'Brien T K, 'Characterisation of delamination onset and growth in a composite laminate', ASTM STP-775, Philadelphia, PA, American Society for Testing and Materials, 1982.

37 Behesty M H, Harris B & Adam T, 'A general fatigue life model for high-performance fibre composites with and without impact damage', *Composites: Part A*, 1999 **30**(8) 971–987.

38 Matthews F L, Kinloch A J, Feng X & Attia O, *Modelling the Fatigue Damage Growth in Fibre-reinforced Plastic Composite Components*, Final Report, Centre for Composite Materials, Imperial College, 1999.

39 Tong J, Guild F J, Ogin S L & Smith P A, 'On matrix crack growth in quasi-isotropic laminates – I: experimental investigation', *Compos Sci Tech*, 1997 **57**(11) 1527–1535.

Index